WIND POWER PLANTS

WIND POWER PLANTS
Theory and Design

by

Désiré LE GOURIÉRÈS

Mechanical Engineer,
Doctor of Science,
Professor of Fluid Mechanics,
University of Dakar
Member of the British Wind Energy Association

PERGAMON PRESS

OXFORD · NEW YORK · TORONTO · SYDNEY · PARIS · FRANKFURT

U.K.	Pergamon Press Ltd., Headington Hill Hall, Oxford OX3 0BW, England
U.S.A.	Pergamon Press Inc., Maxwell House, Fairview Park, Elmsford, New York 10523, U.S.A.
CANADA	Pergamon Press Canada Ltd., Suite 104, 150 Consumers Rd., Willowdale, Ontario M2J 1P9, Canada
AUSTRALIA	Pergamon Press (Aust.) Pty. Ltd., P.O. Box 544, Potts Point, N.S.W. 2011, Australia
FRANCE	Pergamon Press SARL, 24 rue des Ecoles, 75240 Paris, Cedex 05, France
FEDERAL REPUBLIC OF GERMANY	Pergamon Press GmbH, 6242 Kronberg-Taunus, Hammerweg 6, Federal Republic of Germany

First edition 1982

ISBN 0-08-029966-0 (hard cover)
ISBN 0-08-029967-9 (flexicover)

CONTENTS

ACKNOWLEDGMENTS

The author is very grateful to Mr. Lucien Romani, Head of the Eiffel Laboratory in Paris and Inventor of the Nogent Le Roi wind driven generator and to Mr. Louis Vadot, Engineer at the Neyrpic Company in Grenoble, Inventor of the St Remy des Landes wind turbines, for the documentation and the advice, they heve been kind enough to give him.

The author also expresses his thanks to:
— The National Research Council of Canada,
— The U.S. Department of Energy, Wind system branch,
— The American research centres of Amherst, Massachusetts Institute of Technology, and Fort Collins, Colorado (USA),
— The Dutch research centres of Petten and Amersfoort,
— Mr Helge Petersen and the engineers of the Riso National Laboratory in Roskilde, Copenhagen (Denmark),
— The National Swedish Board for Energy source development,
— The Electrical Research Association, the Taylor Woodrow Company and the University of Reading (Great Britain),
— The Growian Company of Hamburg and the research centre of Pellworm (W. Germany),
— Électricité de France and the Neyrpic Company of Grenoble (France), for the photographs and documents that these Companies and Laboratories have sent him.

The present English version has been translated from the French book : « Énergie éolienne, Théorie, Conception et calcul pratique des installations » published by Eyrolles.

The author is particularly grateful to his American, English and Canadian friends for their cooperation in the English translation :
— Dr. Andrew Garrad, Engineer at the Taylor Woodrow Company, Expert in wind energy, who agreed to translate chapter V.
— Professor G.I. Fekete of Mc Gill University in Montreal, for his translation of chapter VI.

— Mr Paul Wenger, Project Development Officer of USAID in Dakar, for having kindly agreed to read the book and offer much constructive advice to the author.

— The members of the American Peace Corps in Dakar.

Without the international assistance of all wind energy researchers, this book could not have been so complete.

The figures and graphs have been drawn by Chérif Coly and the text typed by Fatou Guèye.

The author acknowledges all those who have helped him in his work.

The author wishes to express his special thanks to Dr David Lindley, Manager of Alternative Energy Systems Department at the Taylor Woodrow Construction Limited for writing the Foreword of this book.   David Lindley as manager of the U.K. Wind Energy Group is responsible for the two wind turbines to be erected on Burgar Hill, Orkney.

FOREWORD

The development of Wind Energy Technology has moved at a rapid pace during the last 5 years. There are numerous megawatt rated wind turbines either operating or under construction or in the design phase in several countries including the United States, Denmark, Sweden, Canada, the Netherlands, Germany and the United Kingdom.

Wind Energy, as a renewable energy resource, has been recognised as that nearest to commercial realisation as a major power contributor. Several utilities around the world are either operating or plan to operate wind turbines connected to their power systems. In the United Kingdom, the Department of Energy has recently made it clear that it considers Wind Energy as the number one renewable resource.

The first edition of this book, in French, has been very well received and has now gone into a second edition. The author in preparing this English edition has updated the French text and illustrations considerably to cover recent developments in the technology.

In a technology where few recently written texts exist, this book offers a great deal. The author has stressed the fundamentals, and covered historical and recent developments in the technology around the world quite exhaustively. It must therefore be considered as compulsory reading for all students, designers and users of this technology.

At the present time of acute energy and pollution problems, this book is particularly welcome.

David LINDLEY
Chairman,
British Wind Energy Association

FOREWORD

INTRODUCTION

This book is intended for researchers, engineers and technicians who wish to extend their knowledge in the wind energy field.

It contains the main theories which are indispensable for accurately fixing measurements and characteristics of a wind rotor for producing electricity or pumping water, whatever the type of machine may be : horizontal or vertical-axis. The book, whose main characteristic is simplicity, is laid out according to the following scheme :

After a short introduction and two general chapters, the first relating to the wind, the second to notions of fluid mechanics necessary to the understanding of wind energy problems, the author describes the horizontal-axis installations and, in particular, the various systems of orientation and regulation effectively used.

After having displayed the blade calculations of horizontal-axis systems in chapter IV, the author examines the vertical-axis wind installations in chapter V.

The succeeding chapters deal with pumping water and the production of electricity by wind energy. They contain, among other things, descriptions of small and high power wind plants constructed throughout the world. The problem of adapting the wind rotor to electrical generators or to pumps is studied in its entirety.

Then comes a chapter dealing with applying the above methods of calculation to four projects of various types.

The appendix contains nomograms, aerodynamic characteristics of profiles commonly used, and computer programs for determining rapidly any wind power plant.

To facilitate the understanding of the text, many graphs and photographs (about 200) have been included.

In conclusion, "Wind Power Plants, Theory and Design" tackles all the problems concerning wind energy.

Its is a modern book, new by its contents, well informed, indispensable to all researchers, engineers and manufacturers who are interested in wind energy. Its up-to-date pratical features will be invaluable to those responsible for projects.

THE HISTORY OF WINDMILLS

The conquest of wind energy did not begin yesterday. History teaches us that windmills have existed since the earliest antiquity in Persia, in Iraq, in Egypt and in China. In the seventeenth century B.C. it is said that Hammurabi, king of Babylonia, conceived a plan to irrigate the rich plain of Mesopotamia with the aid of wind energy. The windmills used at that time, in that country, probably included vertical-axis machines similar to those whose ruins remain on the Iranian plateau.

In the third century B.C., in a study dealing with pneumatics, an Egyptian, Hero of Alexandria, designed a four-bladed horizontal axis windmill which provided compressed air to an organ. May we deduce that this kind of windmill was common in Egypt? It is difficult to affirm it. However we can say, without fear of being refuted, that the birth-place of windmotors is to be found in the eastern part of the Mediterranean basin and in China where only vertical axis machines were known.

It was only during the Middle Ages that windmills appeared in Italy, France, Spain, and Portugal. We find them, a little later, in Great Britain, Holland, and Germany. Some authors suggest that their arrival in Europe was due to the Crusaders returning from the Middle East, which is not impossible. The machines used in Europe were horizontal axis windmills with four crossed blades. Their main use was to grind cereals, especially wheat. Holland used them from 1350 A.D. to drain polders. They were then coupled to norias or archimedean screws which could raise water up to five meters. They were also used to extract oil from nuts and grains, to saw wood, to convert old rags into paper, to prepare coloured powders for use as dyes and to make snuff tobacco which was formerly a substitute for cigarettes.

The slow multibladed wind turbine appeared later during the nineteenth century. However, in Leupold Jacob's book "Schauplatz der Wasser Künste" (Journal of Hydraulic Arts), edited in Leipzig in 1724, we already find the design of an eight-bladed self-regulating wind turbine which drives a single acting piston pump by means of a crank shaft and tie rod. Each blade, which is able to pivot round its own axis, is maintained by a spring system so as to be progressively effaced in a high wind and thus the rotor revolves no more quickly in a gale than in a medium wind.

Nevertheless, the multibladed wind turbine did not expand on the old continent but in the United States of America. From 1870, it rapidly conquered the whole country and came back to Europe where it received the name of the "American Windmill".

At the dawn of the 20th century, the first modern fast wind turbines driving electric generators, appeared in France, then spread all over the world. Their invention was due to the French Academician Darrieus.

It is beyond all doubt that, in the past, windmills had known a great success. To mankind, they had furnished the mechanical energy up to then lacking for the achievement of his plans. But with the invention of the steam engine, the internal combustion engine, and the development of electricity, their use was often neglected and abandoned. At the same time, owing to the presence on the market of new means of producing energy, the electric wind generators were not readily accepted. Therefore, the use of wind energy seemed to be more and more forsaken and its very future compromised.

However, history, sometimes, reserves surprises. With the decrease of the world stock of hydrocarbons, the continually increasing demand for energy, and the fear of expanding pollution, wind energy has again come to the fore. It can be used very profitably in windy countries. Machines of 2 000 kW capacity and more have been made and tested, and there are still many other projects in view.

In the present book, we are going to examine the different kinds of machines used in practice and the problems which occur in the wind energy field.

CHAPTER I

✳ THE WIND

Under the influence of the continual atmospheric pressure variations which exist on our planet, air can never be still but, is constantly moving. The resulting air current is the wind. The wind is defined by its direction and its speed.

1. THE WIND DIRECTION

Theoretically, the wind blows from high-pressure zones to low-pressure ones. However at medium and high latitudes, its direction is modified by the earth's rotation. The wind becomes parallel to isobaric lines instead of being perpendicular to them. In the northern hemisphere, the wind revolves counter-clockwise round cyclonic areas and clockwise round anticyclonic areas. In the southern hemisphere, the directions of rotation are reversed.

The wind direction is determined by the direction from which it blows.

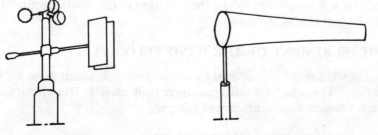

Fig. 1 – *Wind direction and speed indicators.*

Fig. 2 – *Airport wind sock.*

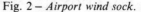

It is a westerly wind if the air current blows from the west.   This direction is given us by the weathercock or the wind vane, a sheet of metal which pivots round a vertical spindle.   The direction may be simply observed from the vane position, in relation to fixed arms pointing to the main points of the compass.

In practice, every reasonably important meteorological station possesses elaborate vanes which constantly record wind speed and direction simultaneously.

The observed data show that the wind direction permanently oscillates around a mean axis.   Using the daily readings taken in each place, a polar diagram can be established showing the percentage of time during which the wind has blown from each direction (numbers written along the radial lines).   The lengths of the radial vectors are chosen so that they are proportional to the mean wind speeds in the reckoned directions.

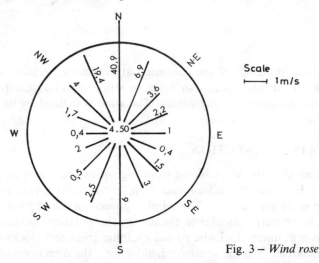

Fig. 3 – *Wind rose*

This kind of diagram which is called a wind rose, can be established for every hour of the day or for every month.   By examining the series of graphs, it is possible to follow the variation of the wind direction during a day or during a year.

2. MEASUREMENT OF THE WIND VELOCITY

The wind speed is measured by anemometers of which there are many varieties.   They can be divided into three main classes : Rotational anemometers, pressure anemometers and others.

a) ROTATIONAL ANEMOMETERS

The most famous models are the cup anemometer of Robinson-Papillon, the Ailleret anemometer made by Cie des compteurs de Montrouge (France) and the Jules Richard anemometer which is a small multibladed windmill.

The first one, *the cup anemometer of Robinson-Papillon*, is found in most of the world's meteorological stations. It is a four-cupped rotor mounted on a short spindle running on ball bearings. At its lower part, the spindle carries a multipole permanent magnet system surrounded by a stator. The indicator measures the voltage which depends on the wind speed. The response time when the windspeed suddenly increases from 10 to 20 m/s, is about 1.3 seconds for the instrument to register a speed of 19 m/s. The cup anemometer starts for 1 to 2 m/s wind speeds.

The second, *the Ailleret anemometer*, has been designed essentially for determining the wind energy. It is possible to link it to a counter or a recorder and thus calculate the recuperable energy. At its base, the anemometer is joined to an alternator whose voltage and frequency are proportional to the wind speed within narrow limits, and influence the coils of a meter. The electrical components of the circuit are adjusted so that the meter integrates the cube of the wind speed. The instrument starts to operate in a wind speed of about 3 m/s.

The Jules Richard anemometer is, above all, a laboratory apparatus and is more sensitive than the cup instrument.

b) PRESSURE ANEMOMETERS

The best known devices are the ancient ball and dial anemometer, the Pitot and Dines instruments, the Era gust anemometers for horizontal and vertical components and the Best Romani anemometer.

The *ball and dial anemometer* is no longer in use. It consists of a ball hung on a thread which moves in front of dial. The angle between the vertical line and the thread depends on the wind velocity.

Another device which is derived from the preceding one, is the inclinable plate anemometer.

The *Pitot and Dines anemometers* are sensitive to the dynamic pressure of the wind. The Dines instrument usually has a wind recorder combined with it so that both the speed and the direction of the wind can be recorded on the same chart.

The *ERA gust anemometers* consist of a ball exposed to wind strength and placed on a vertical or horizontal arm according to whether the horizontal or vertical component is to be measured.

For *the Best Romani anemometer*, the aerodynamic force acts on a vertical cylinder. The direction and the velocity of the wind are determined from recorders using galvanometers. These receive electric currents through resistance gauges placed on metallic blades disposed at the base of the vertical cylinder on the sides of the instrument. The Best Romani

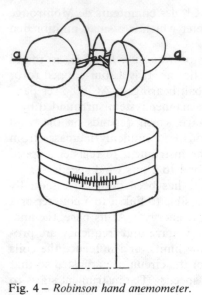

Fig. 4 – *Robinson hand anemometer*.

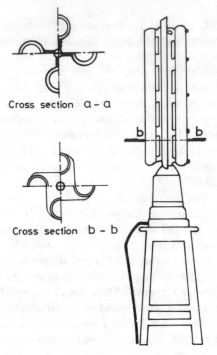

Cross section a – a

Cross section b – b

Fig. 5 – *Ailleret – CdC anemometer*.

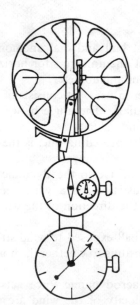

Fig. 6 – *Jules Richard anemometer*.

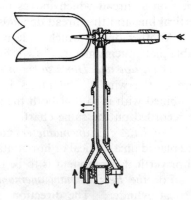

Fig. 7 – *Dines anemometer*.

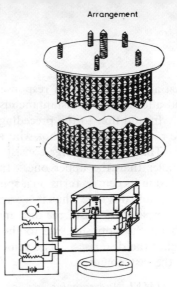

Arrangement

Fig. 8a – *Best-Romani anemometer.*

Fig. 8b – *Principle.*
1. Galvanometer
2. Flexible sheet
3. Gauge

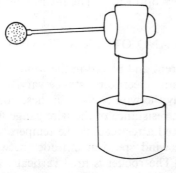

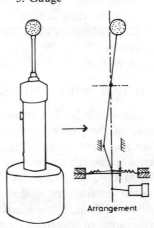

Arrangement

Fig. 9 – *ERA gust anemometer for vertical component.*

Fig. 10 – *ERA gust anemometer for horizontal component.*

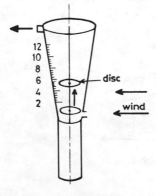

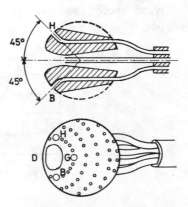

Fig. 11 – *Rotameter.*

Fig. 12 – *I.M.F.L. Anemometer.*

gust anemometer has a response time of less than one-tenth of a second
and allows almost instantaneous recording of wind speeds.

In addition to the preceding apparatus, the IMFL anemoclinometer,
the rotameter and the hot wire anemometer can be used.

The *wind rotameter* consists of a conical tube whose lower section is
smaller than the upper one.   Inside the tube, a plastic disc with a hole
bored in its centre, turns on a spindle.   At the bottom of the tube is a hole.
Air enters this hole when it is oriented to the wind.   Then, it blows through
the apparatus, raises the disc and goes out through an upper orifice situated on the opposite side to the first one.   The higher the velocity, the
higher the disc rises.   The velocity is read from a graduation on the side
of the instrument.

IMFL Anemometer

This instrument has been designed by the Institute of Fluid Engineering
of Lille (France).   It is a spherical gauge with holes bored in the centre and
at 45º to its axis.   These holes are linked to manometers.   These readings
give both the direction and speed of the wind.   The speed is obtained
from the reading on the centre gauge manometer, and the direction from
that relative to the 45º holes diametrically opposite.   The instrument can
be used with up to 40º of inclination.

c) HOT WIRE ANEMOMETERS AND OTHER METHODS

The wire is heated by an electric current and placed in the wind.   The
cooling of the wire by convection makes its electric resistance vary.   The
higher the wind velocity, the more rapidly the wire cools.   In these conditions, it is enough to measure the electric resistance of the wire gauge with
a Wheatstone bridge, to get the wind speed after reading the temperature.

One possible method of measuring wind speed at altitudes between
30 m and 200 m, is *the smoky rocket*.   The rocket is fired vertically and
photographs of the smoky ribbon are taken sideways at regular intervals
(every second, for example).   The analysis of the successive positions
indicates the direction and speed of the wind at the different points.

Wind speed can also be measured with *meteorological sounding
balloons*.

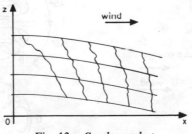

Fig. 13 − *Smoky rocket.*

There are, therefore, many types of anemometers. The most commonly used in the world, is the Robinson anemometer. In practice, the ordinary anemometers associated with summing counters which have the advantage of being economical, are only used for rough studies. To get the necessary fundamental information to erect a complete installation, the preceding instruments are associated with recorders which give the variations of the wind velocity as a function of time.

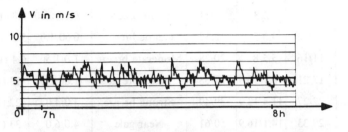

Fig. 14 — *Anemometer record.*

3. WIND VELOCITY

An international scale has been established by Admiral Sir Francis Beaufort which divides wind speeds into 17 strengths or forces. Table 1 gives it, in its latest, revised form.

The last columm gives the average pressure in daN/m^2 on a flat plate perpendicular to the wind for different wind speeds. The relation $p = 0.13\ V^2$ has been used for the evaluation, V being expressed in m/s.

The highest maximal velocities have been measured in tropical hurricanes (speeds of up to 200 km/h are not exceptional), and in the neighbourhood of 45° south latitude known as the roaring forties.

The highest speed ever recorded was at Mount Washington (New Hampshire, USA) April 12th, 1934. During 5 minutes, the mean speed rose to 338 km/h.

4. GENERAL ATMOSPHERIC CIRCULATION

The general atmospheric circulation has its origin in two main phenomena :

Solar radiation,

The earth's rotation.

a) SOLAR RADIATION

Because of the sun's position, the earth is hotter near the equator than near the poles. This causes cold surface winds to blow from the poles to the equator to replace the hot air that rises in the tropics and moves through the upper atmosphere towards the poles.

TABLE 1 : **Beaufort wind scale.**

Beaufort number	Wind speed in			Descriptive terms	Wave height in m	Pressure on a flat plate in daN/m²
	knots	m/s	km/h			
0	1	0/0.4	< 1	Calm	—	—
1	1/3	0.5/1.5	1/6	Light air	—	0.13 (1 m/s)
2	4/5	2/3	7/11	Light breeze	0.15/0.30	0.8 (2.5 m/s)
3	7/10	3.5/5	12/19	Gentle breeze	0.60/1.0	3.2 (5 m/s)
4	11/16	5.5/8	20/28	Moderate breeze	1.0/1.50	6.4 (7 m/s)
5	17/21	8.1/10.9	29/38	Fresh breeze	1.80/2.50	13 (10 m/s)
6	22/27	11.4/13.9	39/49	Strong breeze	3.0/4.0	22 (13 m/s)
7	28/33	14.1/16.9	50/61	Near gale	4.0/6.0	33 (16 m/s)
8	34/40	17.4/20.4	62/74	Gale	5.50/7.50	52 (20 m/s)
9	41/47	20.5/23.9	75/88	Strong gale	7.0/9.75	69 (23 m/s)
10	48/55	24.4/28	89/102	Storm	9.0/12.50	95 (27 m/s)
11	56/63	28.4/32.5	103/117	Violent storm	11.30/16.0	117 (30 m/s)
12	64/71	32.6/35.9	118/133	Hurricane	13.70	160 (35 m/s)
13	72/80	36.9/40.4	134/149			208 (40 m/s)
14	81/89	40.1/45.4	150/166			265 (45 m/s)
15	90/99	45.1/50	167/183			325 (50 m/s)
16	100/108	50.1/54	184/201			365 (54 m/s
17	109/118	54.1/60	202/220			470 (60 m/s)

TABLE 2 : **Gives a description of phenomenas which can be seen on sea and land according to the strength of the wind.**

Beaufort number	Sea Criterion	Land Criterion
0	Sea is like a mirror.	Smoke rises vertically.
1	Ripples with the appearance of scales are formed but without forming crests.	The wind inclines the smoke. But weather-cocks do not rotate.
2	Small wavelets, still short but more pronounced. Crests have a glassy appearance and do not break.	The leaves quiver. One can feel the wind blowing on one's face.

Beaufort number	Sea Criterion	Land Criterion
3	Large wavelets. Crests begin to break. Foam of glassy appearance. Perhaps scattered with white horses.	Leaves and little branches move gently.
4	Small waves, becoming longer : fairly frequent white horses.	The wind blows dust and leaves on to the roads. Branches move.
5	Moderate waves, taking a more pronounced long form; many white horses are formed. (Chance of some spray.)	Little trees begin to sway.
6	Large waves begin to form; the white foam crests are more extensive everywhere. (Probably some spray.)	Big branches move. Electrical wires vibrate. It becomes difficult to use an umbrella.
7	Sea leaps up and white foam from breaking waves begins to be blown in streaks along the direction of the wind.	Trees sway. Walking against the wind becomes unpleasant.
8	Moderately high waves of greater length; edges of crests begin to break into spindrift. The foam is blown in well-marked streaks along the direction of the wind.	Little branches break. It is difficult to walk outside.
9	High waves. Dense streaks of foam along the direction of the wind. Crests of waves begin to topple, tumble and roll over. Spray may affect visibility.	Branches of trees break.
10	Very high waves with long overhanging crests. The resulting foam in great patches is blown in dense white streaks along the direction of the wind. The whole surface of the sea takes on a white appearance. The tumbling of the sea becomes heavy and shocklike. Visibility affected.	Trees are uprooted and roofs are damaged.
11	Exceptionally high waves. (Small and medium sized ships might be lost to view, for a time, behind the waves.) The sea is completely covered with long white patches of foam lying along the direction of the wind. Everywhere the edges of the wave crests are blown into froth. Visibility affected.	Extensive destruction. Roofs are torn off. Houses are destroyed and so on.
12	The air is filled with foam and spray. Sea completely white with driving spray; visibility very seriously affected.	

b) THE EARTH'S ROTATION

The rotation of the earth also affects the atmospheric circulation. Inertia tends to deflect the cold air near the surface of the earth to the west, while the warm air in the upper atmosphere is deflected to the east. This causes large counter-clockwise circulation of air around low pressure areas in the northern hemisphere and clockwise circulation in the southern hemisphere.

In fact, the actuality is more complicated. In practice, atmospheric circulation may be represented as it is shown in fig. 15.

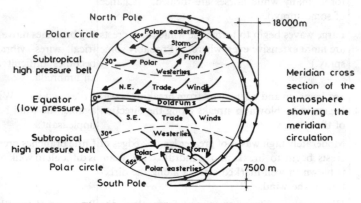

Fig. 15 – *General world wind circulation.*

In each hemisphere, we can discern three more or less individualized cells : a tropical cell, a temperate cell, a polar cell which turn one against the other like cogs in a gear box. The north and south tropical cells are separated from one another by the equatorial calm which is a low pressure area, and from the temperate cells by the subtropical high pressure belts.

Actually, the sketch is not perfect. The unequal heating of oceans and continents, relief, and seasonal variations deform and divide the high and low pressure belts. There are also atmospheric disturbances created by masses of cold air which move, from time to time, from the poles towards the equator. Thus, the state of the atmosphere is continually evolving.

5. GEOGRAPHICAL DISTRIBUTION OF THE WIND ON THE EARTH'S SURFACE

The maps (fig. 16a, 16b, 17a, 17b) give the mean direction and the mean speed of the wind on the earth's surface in January and July. They show that the wind is generally stronger over oceans than over continents. These disparities can be explained by the effects of relief and vegetation which impede the windstream. As a result, the most favourable areas for

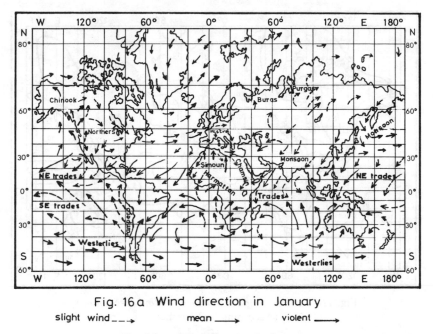

Fig. 16a Wind direction in January
slight wind _ _ →   mean ⎯→   violent ⎯→

Fig. 16a – *Wind direction in January.*

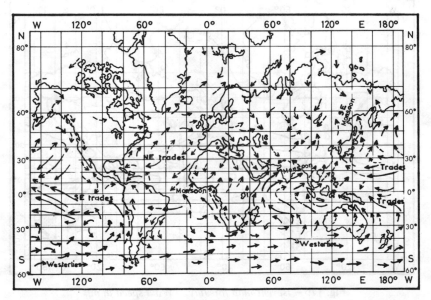

Fig. 16b – *Wind direction in July.*

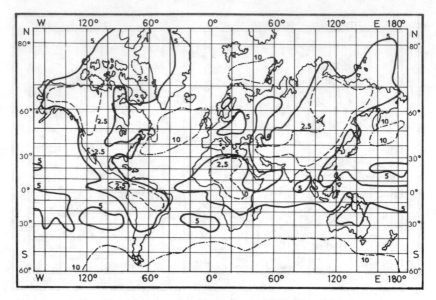

Fig. 17a – *Isovent map for January.   Wind speed in m/s.*
(According to Lauscher)

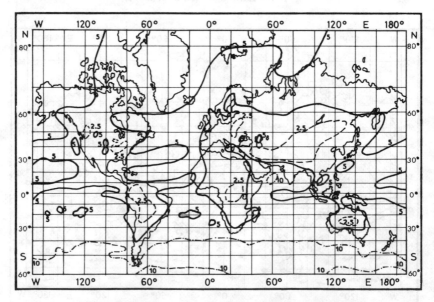

Fig. 17b – *Isovent map for July.   Wind speed in m/s.*

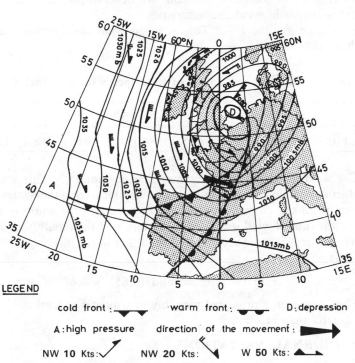

METEOROLOGICAL MAP OF WEST EUROPE
ON JANUARY 19th 1978 AT G. M. T. 12h.

LEGEND

cold front : ◣◣◣◣   warm front : ◖◖◖   D : depression

A : high pressure   direction of the movement : ▬▬▶

NW 10 Kts : ╱   NW 20 Kts : ╲   W 50 Kts : ◢◣▬

Fig. 18 – *The weather map shows that the wind direction is almost parallel to the iso-*
baric lines at high and medium latitudes.

wind energy production are situated on the continents near the seashores.
The best areas are the following :

In Europe : Ireland, Great Britain, France, Netherlands, Scandinavia,
U.S.S.R., Portugal, Greece.

In Africa : Morocco, Mauritania, North West Senegalese coast,
South Africa, Somalia and Madagascar.

In America : South East Coast of Brazil, Argentina, Chile, Canada,
Coastal areas of USA.

In Asia : India, Japan, Coastal areas of China and Vietnam, Siberia.

6. PERIODICAL VARIATIONS OF WIND SPEED

a) SEASONAL VARIATION

As a result of the movement on the earth's surface of the high and low
pressure areas, the speed and direction of the wind generally vary during
the year. The isovent maps for January and July are different as shown
in figures 17a and 17b.

As the positions of cyclonic and anticyclonic areas are repeated every year in relation to the sun's position, we observe periodical and almost similar variations in wind characteristics.

In France, for the Eiffel Tower, mean wind speed varies month by month as table 3 shows.

TABLE 3 : **Monthly coefficients of wind speeds for the Eiffel Tower.**

J	F	M	A	M	J	J	A	S	O	N	D
1.16	1.09	1.06	1.08	0.90	0.85	0.84	0.89	0.88	0.98	1.12	1.15

The wind speed is higher in winter than in summer.   This is a favourable factor because energy needs are also greater during the winter.   This situation is all the more favourable as the energy production is proportional to the cube of wind speed.   The variation of the monthly coefficients relative to energy is more accentuated than that of the monthly coefficients of speed :

$$(1.15)^3 = 1.52 \quad \text{and} \quad (0.85)^3 = 0.61$$

Thus, the energy produced monthly in winter is at least two or three times greater than in summer.

b) DIURNAL VARIATIONS

The wind is subject to diurnal variations due to convective effects.

Because the specific heat of soil is less than that of water, air temperature rises more rapidly during the day over the continents than over the sea.   The hotter air over the land expands, becomes lighter and rises and the cooler, heavier air from over the sea blows in to replace it.   The corresponding inflow is called the sea breeze.

During the night, the direction of the wind is reversed because the land cools more quickly than the water, affecting the air above it.   The continental, cool air blows seawards to replace the warm air that rises from the sea surface. This outflow from land to sea constitutes the land breeze. These breezes may extend up to 50 kilometers from the shore line in medium latitudes and as far as 200 kilometers inland in the tropics.   They may also be observed near lakes.

Similar local breezes occur in mountainous countries.   In the morning, as the summits heat before the valleys, the air over them becomes lighter and rises.   The cooler, heavier air of the valleys moves in to replace it.   Thus during daytime, the wind blows from the valleys towards the mountain tops.   At night time, the air near the summits cools more quickly than the air in the valleys, and the wind direction is reversed.   The relatively cool, heavy air of the summits flows down the slopes into the valleys.

In some countries, the presence of prevailing winds may modify the above mentioned conclusions concerning the direction of the wind. In such cases, the predominant current combines with the breeze to give the actual wind.

As breezes find their origin in thermic phenomena, daily variations in wind direction and velocity are more or less uniform.

7. SUDDEN VARIATIONS OF THE WIND DIRECTION AND VELOCITY

The wind characteristics records show that wind direction and speed are constantly varying. In a very short interval of time, such as one second, the velocity may change from one to two, and the direction may be considerably modified.

Fig. 19a shows a record obtained from an ERA gust anemometer placed at the top of a 10 m tower, at the summit of Costa Hill, Orkney, Great Britain. It refers to measurements of the horizontal component, during a period following the passage of a fast moving cold front. The record shows a change in the wind velocity, from 23 m/s to 37 m/s, in a quarter of a second.

Fig. 19b represents a Best Romani gust anemometer record. This typical record has been extracted from a paper presented by André Argand to the International Conference on New Sources of Energy, in Rome, in 1961. The record gives the variation of both wind direction and velocity.

Fig. 19a – *Gust record from Orkney (U.K.).*

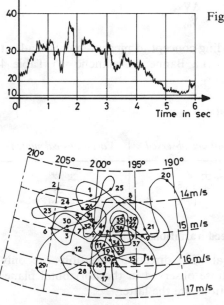

Fig. 19b

Best Romani anemometer record One measure every 0.8 s. Total duration : 32 s. Variation of the direction and the speed of the wind during a gust.

It shows that the wind stream can be considered as the superimposing of a uniform stream on a whirlwind.

A single whirlwind, whose tangential speed is $\Delta \vec{V}$ carried by a uniform stream at the speed $\vec{V}_m$, gives birth to a direction and speed oscillation according to the following rule :

$$\vec{V} = \vec{V}_m + \Delta \vec{V}$$

When the directions of V_m and ΔV are the same, the speed is maximal. When the directions of $\vec{V}_m$ and $\Delta \vec{V}$ are opposite, the speed is minimal. The quotient $\Delta V/V_m$ is usually between 0.3 and 0.4. Assuming that ΔV has a fixed value, we can write :

$$V_{max} = V_m + \Delta V$$
$$V_{min} = V_m - \Delta V$$

whence we can extract :

$$V_m = (V_{max} + V_{min})/2$$
$$\Delta V = (V_{max} - V_{min})/2$$

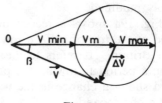

Fig. 20

Let β be the maximal angle between $\vec{V}_m$ and the instantaneous wind speed $\vec{V}$. The amplitude of the direction oscillation is given by the following relation :

$$\sin \beta = \frac{\Delta V}{V_m}$$

L. Vadot has applied the preceding concept to measurements made at the French meteorological station « La Banne d'Ordanche ». Table 4 gives the results he has obtained :

TABLE 4 :

V_{max}	V_{min}	Variations observed	Variations calculated
53 m/s	39 m/s	± 8°	± 9°
36.5 m/s	19.7 m/s	± 22°5	± 24°
20.5 m/s	11 m/s	± 17°8	± 19°

Wind direction and vertical speed variations

Observations show that vertical wind direction variations are only one-tenth to one-fifth as large as those observed in the horizontal plane. Thus the more inconvenient fluctuations are the latter.

8. THE EFFECT OF ALTITUDE

The increase of wind speed with altitude is a well known fact.   Thus in Paris, at the Eiffel Tower, the wind speed varies from 2 m/s at 20 m height to 7-8 m/s at 300 m above ground level.   The reduction of windspeed near the ground is due to friction generated by vegetation, buildings and all sorts of obstacles.

Meteorological data shows that the relative increase of windspeed with altitude may vary from one point to another.   Several authors have proposed to represent the law of variation of wind speed by the following equation :

$$\frac{V}{V_0} = \left(\frac{H}{H_0}\right)^n$$

where V_0 is the observed speed at H_0 meters above the ground topography and V the wind speed at altitude H.

Note that this law is a statistical law which holds good in long series of observations but not necessarily in individual instances.

It is usual to give H_0 the value of 10 m, n being a coefficient varying from 0.10 to 0.40.

The mean velocity profile can also be represented by a log law based on the surface roughness length z_0 of the approach terrain :

$$\frac{V}{V_0} = \frac{Ln(H/z_0)}{Ln(H_0/z_0)}$$

The log law provides the closest fit in the 30-50 m height range but throughout the boundary layer height, the power law is more accurate. Generally because of its simplicity, the power law is more often used.

D. F. Warne and P. C. Calnan have established the following relationship between the surface roughness z_0 and the power-law exponent n :

$$n = 0.04 \, Ln \, z_0 + 0.003 \, (Ln \, z_0)^2 + 0.24$$

Table 5 gives values of z_0 and n for different surface roughnesses.

TABLE 5 : **Values of n and z_0 for different roughnesses at ground surface.**

Terrain type	z_0 in m	n
Smooth (sea, sand, snow)	0.001-0.02	0.10-0.13
Moderately rough (short grass, grass crops, rural areas)	0.02-0.3	0.13-0.20
Rough (woods, suburbs)	0.3-2	0.20-0.27
Very rough (urban areas, tall buildings)	2-10	0.27-0.40

The heights H and H_0 should not be thought of as height above ground surface but above the level of zero wind. This coincides with the average corn height in a dense corn field or with the average wheat height in a wheat field. In a forest, the level of zero wind corresponds to the height where branches of adjacent trees touch.

The energy capable of being intercepted is proportional to V^3. Thus the quotient of energy, at H meters above level ground and at H_0 meters, is given by :

$$\frac{E}{E_0} = \left(\frac{H}{H_0}\right)^{3n} \quad \text{with } 0.30 < 3\,n < 1.20$$

Therefore, to develop maximal power, the wind generator must be mounted as high as possible, for example, for a small wind power plant, on the top of a tower and at least 8 to 10 m higher than the surrounding vegetation or obstacles, in order to avoid high turbulence.

Note that changes in the roughness of the ground affects the wind profile. There is a transition height beyond which the local roughness does not affect the velocity profile. Mr. Duchène-Marullaz (CSTB, Nantes, France) gives, for the transition height $H = 0.08\,x$, x being the distance of the point considered to the change in the ground roughness.

For x > 5 km the transition zone is quite insignificant. For x < 5 km and under the transition height, the characteristics of the wind profiles are related to the roughness of the downwind ground surface and over, to the roughness of the upwind area.

9. THE EFFECTS OF RELIEF: AIRFLOW OVER RIDGES, HILLS AND CLIFFS. CHOICE OF SITE

Orography plays an important role in screening, accelerating and deflecting the wind. Obstacles such as ridges, hills and cliffs affect greatly the wind velocity profile. Some patterns are very favourable for producing wind energy. Others must be avoided because they create considerable turbulence or screening.

Ridges can be defined as elongated hills rising from 100 m or less to 600 m above surrounding terrain and having little or no flat area on their summits. Ridges of hills parallel to seashore with progressive and moderate slopes are very favourable to wind power production, especially, when they are perpendicular to the prevailing wind direction, and when they are bare. Over the summit and in the neighbourhood, the wind is accelerated at very little height and is very regular as shown in fig. 21. The relative increase in speed at the summit can reach 40 %, 60 % or even 80 % for long ridges with progressive slopes. The most suitable slopes range from 10° to 22°.

Note that shoulders (ends) of ridges and the upper half of the windward face are often good sites. But the leeward side must be avoided because of possible turbulence. Ridge sites meriting special consideration are those with features such as pass, gap or saddle.

Similar phenomena occur over circular hills but the increase of the wind speed is lower, and reaches only 20 %, 30 % or 40 % according to the slope of the hill. However, circular hills of moderate slopes can be better sites for wind machines than ridges, if the prevailing wind direction changes very much with the season.

It must be noted that the wind is very accelerated on the sides of the circular hills, tangent to the wind direction. Such places may be considered as good wind power sites, if the prevailing wind direction keeps a nearly constant orientation during the year. But, if this is not the case, the summit is the more suitable place for erecting a wind power installation.

Cliffs with moderate slopes are also very favourable for wind energy production, especially, when they are perpendicular to the wind direction. According to H. Wegley, M. Orgell and R. Drake of the Battelle Institute (USA), the best sites are situated between 0.25 and 2.5 times the cliff height downwind from the cliff. A conservative strategy is to site as close to the cliff edge as possible in order to be sure that the entire rotor rotates well above the zone of turbulence.

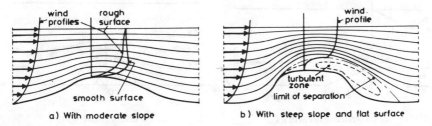

Fig. 21 − *Wind flow over ridges and hills.*

Special care must be taken when choosing a site on a ridge, a hill or a cliff with slopes higher than 30⁰ because on such obstacles, high turbulence may occur. If a hill is very steep, the wind breaks away from the upper surface and considerable turbulence occurs in the wake. Turbulent flows may cause dangerous stresses in the blades of a windmill so that a hill of this kind does not constitute an ideal wind power site. This is also the case for cliffs with steep slopes. Note that whatever the site chosen, it is advisable to make measurements at the positions susceptible to be selected in order to avoid disappointment.

Experimental works

In France, the Electricité de France Company and the CSTB laboratory, in Nantes, have made measurements in wind tunnels concerning the wind flow over ridges and cliffs.

Table 6 gives, for ridges and cliffs of different slopes, the relative increase of the speed with respect to the upstream speed at the same height over the ground surface.

It relates to measurements carried out by C. Sacré. h and z represent respectively the height of the obstacle and the height above ground level. For $z \geqslant 2.5$ h, the speed-up is insignificant.

TABLE 6

| Height over | Zone 1 | | Zone 2 | |
Ground surface	$h/l_1 < 1/3$	$h/l_1 > 1/3$	$h/l_2 < 0.05$	$h/l_2 \geqslant 0.05$
$z < 2.5$ h	$\dfrac{2.5\,h - z}{l_1}$	$\dfrac{2.5\,h - z}{3\,h}$	$\dfrac{2.5\,h - z}{2\,l_1}$	0

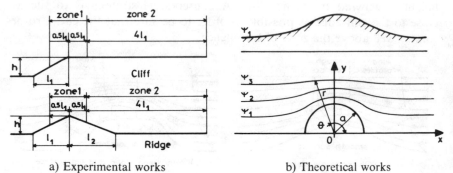

a) Experimental works

b) Theoretical works

Fig. 22 – *Wind over ridges, hills and cliffs.*

Theoretical works

Lamb and Rosenbrock have shown that the streamlines can be defined :

a) over long ridges, by the expression :

$$\psi = U\left(r - \frac{a^2}{r}\right) \sin \theta$$

b) over circular hills, by the relationship :

$$\psi = U\left(r^2 - \frac{a^3}{r}\right) \sin^2 \theta$$

The distribution of velocity vertically above the summits, is given by the following expressions :

— for the case a, by $\quad V = U\left(1 + \dfrac{a^2}{y^2}\right)$

— for the case b, by $\quad V = U\left(1 + \dfrac{a^3}{2y^3}\right)$

In these expressions, U represents the wind velocity up-stream of the obstacle :

a, the radius of the generating cylinder or sphere,
ψ, the constant characterizing a streamline,
y, the height above the point O,
V, the velocity at y metres high,
r, and θ the polar co-ordinates of a point.

But exact hemispheric or hemicylindric hills are in reality very scarce. Fortunately, any streamline can be replaced by a solid surface of the same shape without influencing the stream flow above it. So that, if the ridge (or the circular hill) has a shape which coincides with a streamline, the preceding relationships can be used to calculate the velocity of the wind flow over it.

In fact, the foregoing mathematical expressions do not take into account viscosity forces, so there is some divergence with reality. Simulation in wind tunnels with suitably rough elements or computational techniques give better results.

10. STATISTICAL STUDY OF WIND

From anemometer records and observations, speed duration curves and mean frequency curves may be projected. The drawing of these curves is necessary for any important design.

Annual speed duration curve

On the horizontal axis, the annual time during which the wind velo-

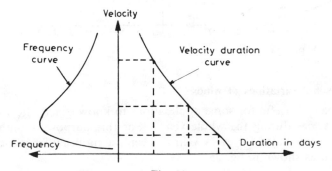

Fig. 23

city exceeds a fixed value, is recorded; that value, itself being given by the other coordinate. We proceed thus year by year. It is then possible to obtain an interannual speed duration curve which will be very useful for energy determination.

Annual frequency curve

This curve can be determined from the preceding one. It gives the annual time during which wind velocity falls within certain limits for instance 3-4 m/s.

This time is equal to the difference of horizontal co-ordinates corresponding to 3 and 4 m/s vertical ordinate of the speed duration curve. The vertical ordinate of the corresponding point of the frequency curve is equal to 3.5 m/s.

Generally, annual frequency curves have only one or two maxima.

Duration of calm spells

The annual duration of calm spells without a break is important because it indicates the period which must be covered by storage when wind-driven plants are used autonomously. Low wind turbines do not operate for pumping water at a speed less than 3 m/s and fast wind turbines at a speed lower than 5 m/s.

Figure 24 indicates the number of periods of a given duration during which the wind velocity was lower than 2, 3 and 5 m/s.

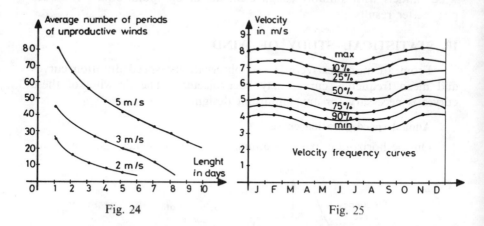

Fig. 24   Fig. 25

Seasonal variations of winds

It may be useful for some applications to know exactly how the wind velocity varies during the whole year. For this purpose, monthly mean speeds are calculated for every year. Then the monthly frequency curves are drawn as shown in fig. 25.

11. THEORETICAL AVAILABLE WIND ENERGY

a) POWER DURATION CURVE

The speed-duration curve enables a curve to be drawn giving the power available as a function of the number of days. This curve may be obtained by cubing the ordinates of the velocity duration curve and keeping the abscissas unaltered. In chapter II, we shall see that the power delivered by an aerogenerator is proportional to the cube of the wind velocity.

So the area situated under the power-duration curve is proportionnal to the available energy in the year :

$$E = \int_0^t P \, dt = K \int_0^t V^3 \, dt$$

The annual energy will be expressed in kWh/m².
We shall see in chapter II (Betz' formula) that K is equal to 0.37.

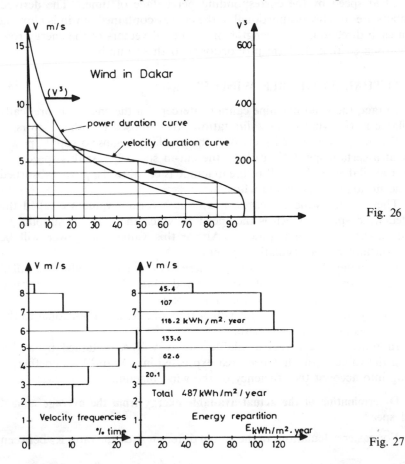

Fig. 26

Fig. 27

b) ENERGY DISTRIBUTION CURVE

We can divide the whole scale of the velocity values into different intervals of 1 m/s.   For every interval of speed, it is possible to evaluate the available energy from the velocity frequency curve.

The energy distribution curve is obtained by cubing the mean speed relative to the interval of velocity and multiplying the result by the corresponding ordinate of the speed frequency curve (number of hours in the year during which the considered velocities occur).

c) WIND ENERGY DISTRIBUTION AS A FUNCTION OF THE WIND DIRECTION: THE ENERGY ROSE

This diagram gives the available energy for each direction of the compass.   The energy rose can be constructed from the wind rose graph by first cubing the average wind speed for each direction then multiplying the cubed speed by the corresponding percentage of time.   The derived numbers are roughly proportional to the energy contained in winds blowing from each direction.   The lengths of the radial vectors of the energy rose are chosen so that they are proportional to these numbers.

12. ACTUAL AVAILABLE WIND ENERGY

In fact, the wind machine cannot intercept all the energy theoretically available in the wind because limitations are imposed by other factors.

When the wind speed increases, the machine begins to supply power only at a certain speed V_m called "the cut-in speed".   At this speed, the power available on the shaft of the machine is equal to the power absorbed by the no-load losses of the whole set.

Then, as the wind increases, it attains a certain value V_N called the "rated wind speed" which is the speed at which the machine supplies its rated power or nominal power.   Above this value, the power will be kept constant by the regulating system.

If the wind velocity continues to increase, it reaches a value V_M called the "furling speed" or cut-out speed beyond which the machine is stopped for reasons of safety, and gives no more power.

Finally, the effective energy, which is available, is proportional to the hatched area as shown in fig. 28 b.

In practice, the actual available energy will be obtained by multiplying the value of this hatched area expressed in kWh/m^2 by a coefficient taking into account the efficiency of the wind machine.

Determination of the actual available energy from the average annual wind speed

For easier calculations of the actual different sites, the ratio K_u between

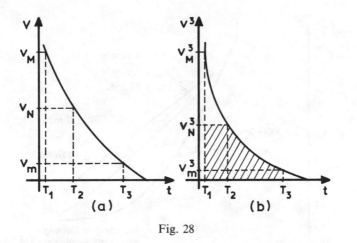

Fig. 28

the actual available energy produced by a given machine and the energy obtained by considering wind speed as a constant equal to the average annual value may be used.

This coefficient which is known as the "usable energy pattern factor", is given by the expression:

$$K_u = \frac{V_N^3(T_2 - T_1) + \int_{T_2}^{T_3} V^3 \, dt}{\overline{V}^3 T}$$

where V is the instantaneous wind speed;

$\overline{V}$, the mean annual wind speed $\left(\overline{V} = \frac{1}{T} \int_0^T V \, dt\right)$

V_N, the rated speed;

T_1, the annual time during which the value of the speed is above the furling speed V_M;

T_2, the annual time during which the speed is greater than the rated speed V_N;

T_3, the annual time during which the speed is greater than the cut-in speed V_m.

The variation of the coefficient K_u as a function of mean annual windspeed has been studied by E. W. Golding for different meteorological stations of West Europe and for different values of rated speeds. The results are shown in fig. 29.

The knowledge of the coefficient K_u enables the calculation of the available energy in sites where only summing counters of wind speed are used. However, the application of the method needs the preliminary determination of the ratio K_u from anemometer records in areas subjected to winds of same origin and comparable velocities.

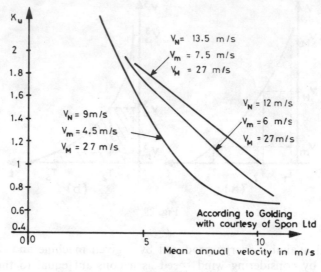

Fig. 29

Figures 30a, 30b, 31a, 31b, give respectively :

— the mean wind speeds at 10 m height, above ground level in Western Europe ;

— the annual available wind energy at 10 m height, above exposed areas of France, Great Britain and Ireland. (At 40 m elevation, the energy is about 2 or 3 times higher than at 10 m).   In fact, no more than 50 % approximately, can be intercepted ;

— the annual available wind energy at 50 m height above ground level in the USA and Canada ;

— the annual available energy available in different parts of the world in kWh/year per rated kilowatt output for wind machines designed for rated wind speeds of 11.1 m/s.

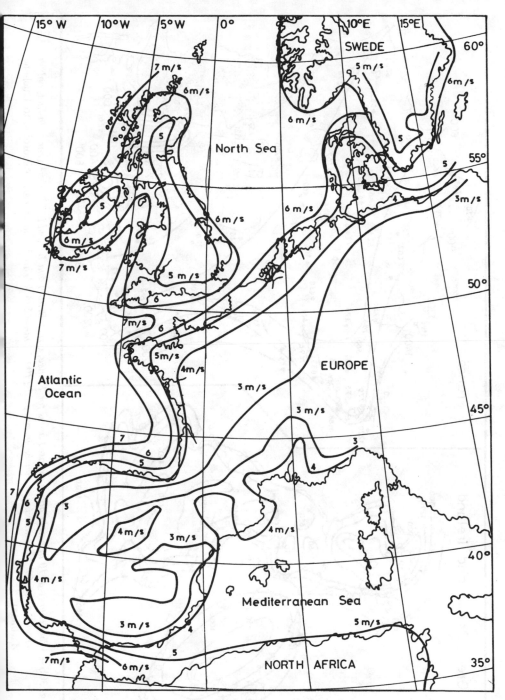

Fig. 30a – *Mean annual wind speeds in western Europe*
(in m/s at 10 m height above ground surface).

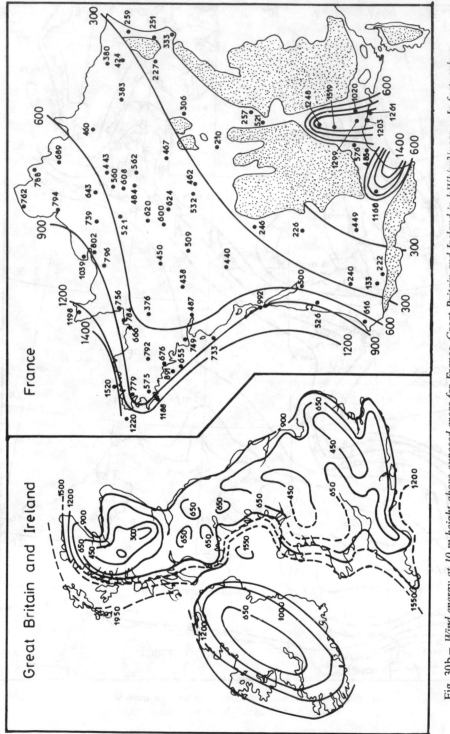

Fig. 30b – *Wind energy at 10 m height above exposed areas for France, Great Britain and Ireland in kWh/m²/year. In fact, only a percentage of 50 % at maximum can be intercepted (100 W/m² = 876 kWh/m²/year ≃ 900) (According to Duchêne Marullaz).*

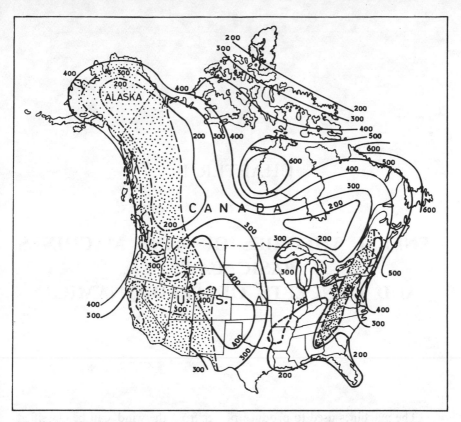

Fig. 31a – *Wind energy at 50 m above exposed areas for Canada and the United States in W/m² (100 W/m² = 876 kWh/m²/year).*

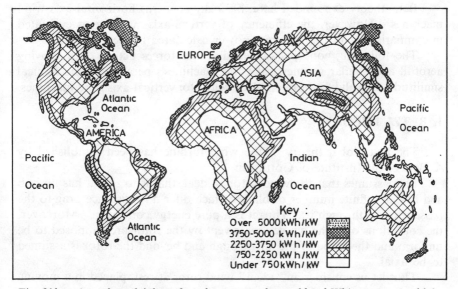

Key:
Over 5000 kWh/kW
3750-5000 kWh/kW
2250-3750 kWh/kW
750-2250 kWh/kW
Under 750 kWh/kW

Fig. 31b – *Annual availability of wind energy in the world in kWh/year per rated kilowatt output for wind machines designed for rated wind speeds of 11 m/s (According to Frank Eldridge).*

CHAPTER II

GENERAL THEORIES OF WIND MACHINES
BASIC LAWS
AND CONCEPTS OF AERODYNAMICS

The machines used to produce power from the wind, can be classified mainly in two different groups : horizontal-axis machines and vertical-axis machines.

Betz' theory expounded here, basically concerns horizontal-axis wind machines. However, the efficiency of vertical-axis machines is evaluated in comparison with the maximal power calculated by Betz' formula.

The following concepts of aerodynamic forces acting on a moving aerofoil and similar geometrical wind machines operating in mechanical similitude, are valid for horizontal as well as for vertical-axis wind machines.

1. BETZ' THEORY

The first global theory of the wind turbine has been established by A. Betz of the Institute of Göttingen.

Betz assumes that the wind rotor is ideal, that is to say, it has no hub and has an infinite number of blades which offer no resistance drag to the passage of air through it. Thus it is a pure energy converter. Moreover, the conditions over the whole area swept by the rotor are supposed to be uniform and the speed of the air through and beyond the rotor is assumed to be axial.

Thus let us consider an "ideal" wind rotor at rest, placed in a moving atmosphere.

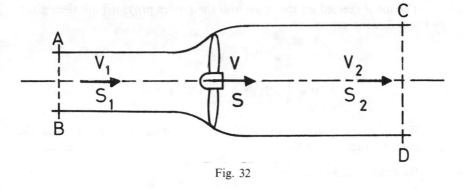

Fig. 32

Let V_1 be the wind speed at a considerable distance upwind,

V, the wind speed actually passing through the rotor and assumed to be uniform over the whole area S swept by the blades,

V_2, the wind speed downwind, far from the rotor.

The section of the air flow which passes through the rotor is S_1 upwind of the rotor and S_2 downwind.

The production of mechanical energy by the rotor is possible only by reducing the kinetic energy of the air. Thus V_2 is necessarily lower than V_1. Consequently, the section of the air flow which passes through the rotor increases from upstream to downstream. S_2 is greater than S_1.

If we suppose that the air is incompressible, the continuity condition (constant mass flow) can be written as :

$$S_1 V_1 = S V = S_2 V_2$$

The force exerted on the wind rotor by the wind is given by Euler's theorem and is equal to :

$$F = \rho S V (V_1 - V_2)$$

The power absorbed is therefore :

$$P = F V = \rho S V^2 (V_1 - V_2)$$

As we have said, this power has been taken from the kinetic energy.

The variation of the kinetic energy from upstream to downstream amounts to :

$$\Delta T = \frac{1}{2} \rho S V (V_1^2 - V_2^2)$$

Equating the two expressions P and ΔT, we obtain :

$$V = \frac{V_1 + V_2}{2}$$

The force exerted on the rotor and the power provided are then given by the following expressions :

$$F = \frac{1}{2} \rho S (V_1^2 - V_2^2)$$

$$P = \frac{1}{4} \rho S (V_1^2 - V_2^2) (V_1 + V_2)$$

For a given upstream speed V_1, we can study the variation of the power P as a function of V_2.

By differentiation, we get :

$$\frac{dP}{dV_2} = \frac{1}{4} \rho S (V_1^2 - 2V_1 V_2 - 3V_2^2)$$

The equation $\frac{dP}{dV_2} = 0$ has two solutions :

— the first one   : $V_2 = - V_1$ which has no physical meaning,
— the second one : $V_2 = \frac{V_1}{3}$ which corresponds to the maximum power.

On substituting the particular value $V_2 = V_1/3$ in the expression of P, we obtain the maximum power which can be produced :

$$P_{max} = \frac{8}{27} \rho S V_1^3 = 0.37 \ S V_1^3$$

taking for ρ, specific mass of air : 1.25 kg/m³.

This expression constitutes Betz' formula.

It supposes the direction of wind velocity through the rotor is axial, and that the velocity is uniform over the area S.

2. WING AND AEROFOIL : GEOMETRY AND AERODYNAMIC CHARACTERISTICS

The main element of the wind machine, whatever the type may be (horizontal or vertical-axis machine) is the blade.   This may be considered as a rotating wing.   To have a good understanding of its action and especially to choose its optimal shape and dimensions, it is necessary to have some basic knowledge of the aerodynamics of aerofoils.   The Betz' formula does not give us the method to construct blades.

Therefore consider a motionless aerofoil subject to a wind of velocity $\vec{V}$. Let us give some definitions concerning its geometry and its position relative to the speed vector $\vec{V}$ assumed to be parallel to the section of the aerofoil.

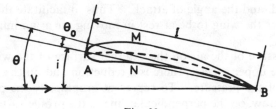

Fig. 33

a) DEFINITIONS

The sharp end of the profile (point B) is called the "trailing edge". The "leading edge" is the locus of the point A of the nose of the profile which is the farthest from the trailing edge.

AB = 1 is the chord of the profile,
AMB, the upper surface,
ANB, the lower surface.

At any distance along the chord from the nose, a point may be marked mid-way between the upper and lower surfaces. The locus of all such points, usually curved, is the camber line or median line of the section.

The incidence angle is the angle i between the chord and the air speed vector $\vec{V}$ at infinite upstream.

The zero lift angle is the angle θ_0 between the chord and the zero lift line.

The lift angle is the angle θ between the zero lift line and the air speed vector $\vec{V}$ at infinite upstream.

$$i = \theta + \theta_0$$
$$\theta = i - \theta_0$$

Here, θ_0 is negative, θ and i are positive.

Another parameter is the maximum thickness h. This, when expressed as a fraction of the chord, is called the thickness/chord ratio or relative thickness. Current values in use range from 3 % to 20 % (10 % to 15 % for the usual wind rotors). The position along the chord at which the maximum thickness occurs may vary between 20 % and 60 % of the chord from the leading edge (usually around 30 % for the aerofoils of wind rotors).

b) AERODYNAMIC FORCE EXERTED ON A MOVING WING IN A STILL ATMOSPHERE

If we suppose the wing to be at rest and the air to be moving at the same speed, but in the opposite direction, the aerodynamic force exerted on the wing does not change in value. The effort exerted only depends on the relative speed, and the angle of attack. Thus to facilitate the explanation, let us consider the wing to be at rest in moving air at an infinite upstream speed V.

The pressure of the air on the external surface of the wing is not uniform : On the upper surface, there is a reduction and on the lower surface, an increase of the pressure. To represent graphically the pressure variations, let us draw, on the perpendicular line to the profile surface, a segment whose length is equal to K_p :

$$K_p = \frac{p - p_0}{\frac{1}{2}\rho V^2}$$

where p is the static pressure at the base of the perpendicular line on the surface, and ρ, p_0, V the conditions at infinite upstream, that is the conditions in the undisturbed flow far from the profiled section.

Joining the tips of the various segments K_p, we obtain the curved line represented fig. 34. K_p is negative for the points of the upper surface and positive for the lower surface.

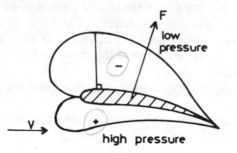

Fig. 34

The resultant of the different elementary forces acting on the wing is a force F, generally inclined with respect to the relative speed direction, and given by the expression :

$$F = \frac{1}{2}\rho C_r S V^2$$

where : ρ is the specific mass of the air
         S the area equal to the product chord $\times$ length of the wing
         C_r the total aerodynamic coefficient.
This force can be divided into two components :
— a component parallel to vector $\vec{V}$: the drag $\vec{F_d}$
— a component perpendicular to vector $\vec{V}$: the lift F_l
F_d and F_l are given by the expressions :

$$F_d = \frac{1}{2}\rho C_d S V^2 \qquad F_l = \frac{1}{2}\rho C_l S V^2$$

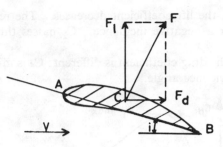

Fig. 35

where C_d and C_l are respectively the drag and the lift coefficients. These components being perpendicular, we can write :

$$F_d{}^2 + F_l{}^2 = F^2$$

and therefore :

$$C_d{}^2 + C_l{}^2 = C_r{}^2$$

Let M be the aerodynamic moment of F relative to the leading edge. We can define a pitching moment coefficient C_m by the relation :

$$M = \frac{1}{2} \rho C_m S l V^2$$

where l is the chord.

Hence the aerodynamic forces on the aerofoil section may be represented by a lift, a drag and a pitching moment. Now, at each value of incidence angle, there will be one particular point C about which the moment of the aerodynamic force F is zero. This special point is termed the centre of pressure. The aerodynamic effects on the aerofoil section may be represented by the lift and the drag alone acting at that point. The position of the centre of pressure relative to the leading edge is determined by the ratio :

$$CP = \frac{AC}{AB} = \frac{x_1}{l} = \frac{C_m}{C_l}$$

Usually $CP = 25\% - 30\%$.

c) VARIATION OF THE LIFT AND DRAG AERODYNAMIC COEFFICIENT

1°) Variation of C_l and C_d versus incidence

The variation is illustrated in fig. 36.

Considering first the variation of the lift coefficient, the representative curve is seen to consist of a straight line curving over at the higher value of C_1 max, at an incidence i_M known as the stalling point.

After the stalling point, the lift coefficient decreases. The representative line is also curved for a negative incidence. C_1 passes through a minimum value C_1 min.

The variation curve of the drag coefficient is different: C_d is minimum for a certain value of the incidence angle.

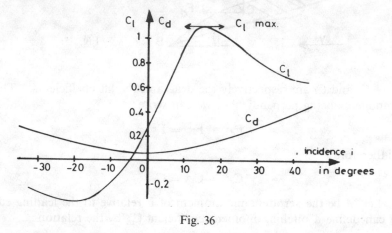

Fig. 36

2°) Lift coefficient versus drag coefficient (Eiffel Polar)

The variation of the lift coefficient versus the drag coefficient is shown in figure 37.

The slope of the straight line OM is: $\tan \theta = C_l/C_d$.

When OM is tangent to the C_l/C_d curve, $\tan \theta$ is maximum and C_d/C_l minimum.

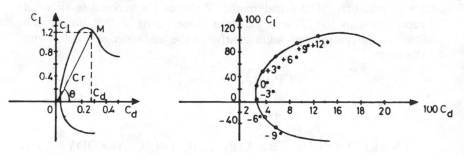

Fig. 37 – *The Eiffel polar.*

The curve of variation is usually graduated in values of incidence angle.

Remark 1 : According to the Bernouilli theorem, the velocity of the stream above the wing is higher, and that below the wing, is lower than the velocity of the undisturbed flow, far from the profile. Thus, the flow around the aerofoil can be regarded as the combination of two different types of flow. The first one is the normal flow around the aerofoil at zero lift when the aerofoil is placed in a uniform stream. The other one is a flow in which the air circulates around the aerofoil forwards over the lower surface, and backwards, over the upper surface. The lift of the aerofoil is associated with the latter.

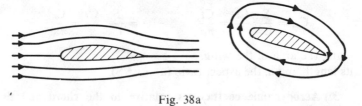

Fig. 38a

Remark 2 : The above result applies only to wings having an infinite length. For a wing whose length is limited, the results must be corrected. The pressure on the lower surface of a lifting wing is greater than that of the surrounding atmosphere while the pressure on the upper surface is lower. Thus, at the tips, air tends to flow from the lower surface towards the upper surface. The result is the creation of vortices at the tips of the wing. In fact, many small vortices appear all along the wing because of the influence of the tips. These small vortices roll up in two large vortices just inboard of the wing tips.

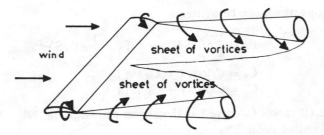

Fig. 38b – *The horseshoe vortex.*

The consequence of the creation of these vortices is an increase of the drag. An induced drag F_{di} occurs, which adds to the preceding one.

$$F_{di} = \frac{1}{2} \rho C_{di} S V^2$$

Thus the drag coefficient becomes :

$$C_d = C_{d0} + C_{di}$$

where C_{d0} is the drag coefficient for a wing of infinite length.

Furthermore, to get the same lift, the angle of incidence must be increased by a quantity ϕ. Thus the new incidence angle to obtain the same lift is:

$$i = i_0 + \phi$$

In fluid mechanics, when the repartition of the circulation is elliptic, it may be shown that C_{di} and ϕ are given by the following relationships:

$$C_{di} = \frac{S}{L^2} \cdot \frac{C_l^2}{\pi} = \frac{C_l^2}{\pi a} \qquad \phi = \frac{S}{L^2} \cdot \frac{C_l}{\pi} = \frac{C_l}{\pi a}$$

S being the area of the wing,
L its length and a, the aspect ratio $(a = L^2/S)$.

3°) Aerodynamic coefficients relative to the chord and the perpendicular (Lilienthal polar)

If we project F on the chord and on the perpendicular to the chord (fig. 39a), we arrive at the following components:

on the chord:

$$F_t = \frac{1}{2}\rho S V^2 (C_d \cos i - C_l \sin i).$$

on the perpendicular

$$F_n = \frac{1}{2}\rho S V^2 (C_l \cos i + C_d \sin i)$$

expressions which can be written as:

$$F_t = \frac{1}{2}\rho C_t S V^2 \text{ and } F_n = \frac{1}{2}\rho C_n S V^2$$

where:
$$C_n = C_l \cos i + C_d \sin i$$
$$C_t = C_d \cos i - C_l \sin i$$

The C_n versus C_t curve has the aspect shown in fig. 39b. It is called the "Lilienthal polar".

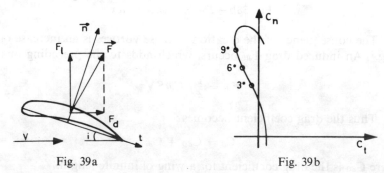

Fig. 39a   Fig. 39b

In practice, this curve is used to determine the thickness of the aerofoil in order to resist the aerodynamic efforts.

3. AERODYNAMICS OF THE ROTOR

a) GEOMETRICAL DEFINITIONS

Ancient windmills or modern wind turbines have many blades which are fastened on a hub and constitute the rotor. Before undertaking their study, let us give some definitions. It is usual to call :

— *rotor axis :* the axis of rotation of the rotor,

— *plane of rotation :* the plane perpendicular to the rotor axis in which the blades revolve,

— *rotor diameter :* the diameter of the area swept by the rotor,

— *blade axis :* the longitudinal axis around which it is possible to make the inclination of the blade vary relative to the plane of rotation,

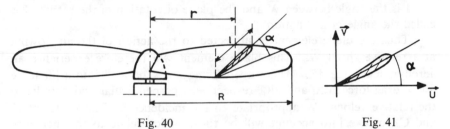

Fig. 40   Fig. 41

— *blade section at radius r :* the intersection of the blade by a cylinder of radius r and whose axis is the rotor axis,

— *setting angle or pitch angle :* the angle α between the chord of the aerofoil section at radius r and the plane of rotation,

— *geometric pitch of the blade section at radius r :* the pitch of the geometric helix of radius r having for axis, the rotor axis, and which is tangent to the chord at radius r. Let α be the setting angle at radius r, the geometric pitch of the section is :

$$H = 2\pi\, r\, \tan\alpha$$

For an airscrew having a constant geometric pitch, the quantity r tan α remains constant along the chord.

b) PERFORMANCES OF A BLADE ELEMENT (ELEMENTARY THEORY)

Consider an element of length dr, chord 1 and pitching angle α at radius r of a rotor blade.

This element has a speed in the plane of rotation equal to $U = 2\pi r N$.

If we call V the axial speed of the wind through the rotor, then the velocity of the air flow relative to the blade is $\vec{W}$ as shown in fig. 42.

$$\vec{V} = \vec{U} + \vec{W} \qquad \vec{W} = \vec{V} - \vec{U}$$

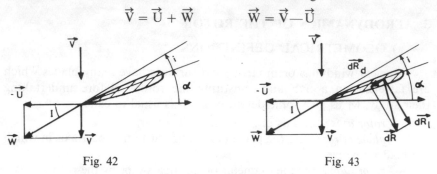

Fig. 42   Fig. 43

The incidence angle is $i = I - \alpha$.

I is the angle between $\vec{W}$ and the plane of rotation of the rotor. I is called the angle of inclination.

Thus the blade element is subjected to the action of the air flowing at a relative speed $\vec{W}$. This blade element will therefore experience an aerodynamic force dR. This force dR may be separated into a lift force and a drag force: dR_l and dR_d, respectively, perpendicular and parallel to the relative velocity $\vec{W}$ appropriate to the incidence i. The values of C_l and C_d to take into account will be those which relate to the incidence angle i for the profile used in the blade element construction.

Estimate the contribution of the aerodynamic force dR into the axial thrust exerted by the wind on the rotor and into the torque which acts on the rotor axis.

Let these contributions be dF and dM.

dF is the projection of dR on the rotor axis and dM, the moment relative to the rotor axis of the projection of dR on the plane of rotation.

$$dF = dR_l \cos I + dR_d \sin I$$
$$dM = r(dR_l \sin I - dR_d \cos I)$$

If we take into account the relationships:

$$dR_l = \frac{1}{2} \rho C_l W^2 dS \text{ and } dR_d = \frac{1}{2} \rho C_d W^2 dS$$

$$W^2 = V^2 + U^2 = V^2 + \omega^2 r^2 \qquad \omega r = \cotg I$$
$$dP = \omega dM$$

we get for dF, dM and dP the following expressions:

$$dF = \frac{1}{2} \rho V^2 dS (1 + \cot^2 I)(C_l \cos I + C_d \sin I)$$

$$dM = \frac{1}{2}\, \rho\, V^2\, rdS\, (1 + \cot^2 I)\, (C_l \sin I - C_d \cos I)$$

$$dP = \frac{1}{2}\, \rho\, V^3\, dS\, \cot I\, (1 + \cot^2 I)\, (C_l \sin I - C_d \cos I)$$

c) GENERAL EXPRESSIONS FOR THRUST, TORQUE AND POWER

The total thrust F exerted by the wind on the rotor and the torque M which acts on the rotor shaft are obtained by adding, respectively, all the elementary forces dF and the elementary moments dM which act on the blades. Therefore, the power P transmitted by the wind to the rotor and the useful power P_u provided by the wind turbine may be calculated without difficulty, for different conditions, by the expressions:

$$P = \Sigma\, dF \cdot V = FV$$
$$P_u = M\omega$$

The efficiency is given by:

$$\eta = \frac{P_u}{P} = \frac{M\omega}{FV}$$

4. PERFORMANCES OF SIMILAR GEOMETRIC WIND MACHINES

In the preceding paragraph, we have briefly analysed the aerodynamic forces acting on the rotor, but we have neglected some factors such as the interaction between blades. To get a more accurate idea of the influence of such parameters, it is necessary to do experiments on physical models in a wind tunnel.

The geometry of the model is similar to that of the prototype we intend to build. The model is wind tested in conditions of kinematic similarity, that is to say: the incidence angles of the air on each corresponding blade element of the model and the prototype are the same.

These conditions mean that the following ratios must be equal:

$$\frac{V_{11}}{V_{12}} = \frac{V_{21}}{V_{22}} = \frac{V_1}{V_2} = \frac{W_{10}}{W_{20}} = \frac{U_{10}}{U_{20}}$$

The number 1 is relative to the prototype and the number 2 to the model.

V_{11} and V_{12} are the corresponding speeds upstream of the prototype and the model,

V_{21} and V_{22}, the downstream speeds,

V_1 and V_2, the wind speed through the wind rotors,

U_{10} and U_{20} the circumferential tip speeds in the plane of rotation,

U_1 and U_2, the circumferential speed of corresponding elements of the prototype and the model.

The relationships :

$$\frac{V_1}{V_2} = \frac{U_{10}}{U_{20}} = \frac{V_{11}}{V_{12}} \quad \text{can be written as :}$$

$$\frac{U_{10}}{V_1} = \frac{U_{20}}{V_2} \quad \text{and} \quad \frac{U_{10}}{V_{11}} = \frac{U_{20}}{V_{12}}$$

The first expression leads us to conclude that the angles of inclination I_0 are equal at the blade tips of the prototype and the model; the second one, that the tip-speed ratio λ_0 (quotient of the circumferential speed at the blade tip by the upstream wind speed) must be the same for the prototype and the model.

It should be noted that, if the inclination angles I are equal at the blade tips of the prototype and the model, they are also equal at corresponding radii.

In fact, if we call I_1 and I_2 the angles of inclination in the sections, we can write :

$$\cot I_1 = \frac{U_1}{V_1} = \frac{\omega_1 r_1}{V_1}$$

$$\cot I_2 = \frac{U_2}{V_2} = \frac{\omega_2 r_2}{V_2}$$

Taking into account the relations :

$$\frac{V_1}{V_2} = \frac{U_{10}}{U_{20}} = \frac{\omega_1 R_1}{\omega_2 R_2} = \frac{\omega_1 r_1}{\omega_2 r_2}$$

and dividing the preceding equations, we get :

$$\frac{\cot I_2}{\cot I_1} = \frac{\omega_2 r_2}{\omega_1 r_1} \times \frac{V_1}{V_2} = 1$$

Hence we conclude that the angles of inclination I are equal for the corresponding elements of the prototype and the model.

The setting angles being also equal, the incidence angles which are equal to the difference ($i = I - \alpha$) are accordingly the same. Thus C_1 and C_d have the same values.

Referring to the expressions of the thrust, torque and power produced by the blade element :

$$dF = \frac{1}{2} \rho V^2 \, dS \, (1 + \cot^2 I) \, (C_1 \cos I + C_d \sin I)$$

$$dM = \frac{1}{2} \rho V^2 \, rdS \, (1 + \cot^2 I) \, (C_1 \sin I - C_d \cos I)$$

$$dP = \frac{1}{2} \rho V^3 \, dS \cot I \, (1 + \cot^2 I) \, (C_1 \sin I - C_d \cos I)$$

and dividing dF by $\rho V^2 dS$, dM by $\rho V^2 r dS$, dP by $\rho V^3 dS$, we obtain expressions which contain only in their second member cot I, C_l and C_d. We have shown that the values of I, i, C_l, C_d are the same for the corresponding elements of the prototype and the model when conditions of similarity are fulfilled. Therefore, for the corresponding blade elements of the prototype and the model, the following ratios are equal:

$$\frac{dF_1}{\rho_1 V_1^2 dS_1} = \frac{dF_2}{\rho_2 V_2^2 dS_2} \quad , \quad dF_1 = dF_2 \frac{\rho_1 V_1^2 dS_1}{\rho_2 V_2^2 dS_2} = dF_2 \frac{\rho_1 V_1^2 D_1^2}{\rho_2 V_2^2 D_1^2}$$

$$\frac{dM_1}{\rho_1 V_1^2 r_1 dS_1} = \frac{dM_2}{\rho_2 V_2^2 r_2 dS_2} , \quad dM_1 = dM_2 \frac{\rho_1 V_1^2 r_1 dS_1}{\rho_2 V_2^2 r_2 dS_2} = dM_2 \frac{\rho_1 V_1^2 D_1^3}{\rho_2 V_2^2 D_2^3}$$

$$\frac{dP_1}{\rho_1 V_1^3 dS_1} = \frac{dP_2}{\rho_2 V_2^3 dS_2} \quad , \quad dP_1 = dP_2 \frac{\rho_1 V_1^3 D_1^2}{\rho_2 V_2^3 D_2^2}$$

As we have said, the total thrust, the torque and the power are obtained by adding the elementary efforts, the elementary torques and the elementary powers.

$$F_1 = \Sigma dF_1 = \frac{\rho_1 V_1^2 D_1^2}{\rho_2 V_2^2 D_2^2} \Sigma dF_2 = \frac{\rho_1 V_1^2 D_1^2}{\rho_2 V_2^2 D_2^2} F_2$$

This can be written as:

$$\frac{F_1}{\rho_1 V_1^2 D_1^2} = \frac{F_2}{\rho_2 V_2^2 D_2^2}$$

In the same manner:

$$M_1 = \Sigma dM_1 = \frac{\rho_1 V_1^2 D_1^3}{\rho_2 V_2^2 D_2^3} \Sigma dM_2 = \frac{\rho_1 V_1^2 D_1^3}{\rho_2 V_2^2 D_2^3} M_2$$

$$\frac{M_1}{\rho_1 V_1^2 D_1^3} = \frac{M_2}{\rho_2 V_2^2 D_2^3}$$

and

$$P_1 = \Sigma dP_1 = \frac{\rho_1 V_1^3 D_1^2}{\rho_2 V_2^3 D_2^2} \Sigma dP_2 = \frac{\rho_1 V_1^3 D_1^2}{\rho_2 V_2^3 D_2^2} P_2$$

$$\frac{P_1}{\rho_1 V_1^3 D_1^2} = \frac{P_2}{\rho_2 V_2^3 D_2^2}$$

The preceding equations suppose that the conditions of similitude are fulfilled.

The last one shows that the efficiency of the prototype and that of the model are equal for the same tip-speed ratios. The denominators of the last fractions are effectively proportional to the kinetic energy of the air flow which passes through the prototype and the model rotors. One can see the great advantage of this result, which enables us to determine the efficiency of a big machine from the tests done on a similar geometric small one, tested in a wind tunnel.

The preceding expressions establish relationships between the wind speed through the rotor, the thrust, the torque and the power of the prototype and the model. These expressions are still true, if we substitute the wind speed V_{11} and V_{12} upstream of the wind machine for the wind speeds V_1 and V_2 through the rotor as the denomitor. As a matter of fact, these speeds are connected by the relationship:

$$\frac{V_1}{V_2} = \frac{V_{11}}{V_{21}}$$

Thus the performances of wind machines will be evaluated relative to the upstream wind speed because, there, the wind is undisturbed.

In practice, one represents the variation curve of the quantities:

$$C_F = \frac{2F}{\rho S V^2} \qquad C_m = \frac{2M}{\rho S V^2 R} \qquad C_p = \frac{2P}{\rho S V^3}$$

versus the tip-speed ratio $\lambda_0 = \dfrac{U_0}{V}$

V, being the speed of the wind upstream of the wind machine at five or six diameters in front of the turbine, U_0, the circumferential speed at the blade tip and S, the swept area.

C_F, C_m and C_p are called respectively axial thrust coefficient, moment coefficient and power coefficient.

The advantages of such a graphical representation are obvious: the characteristic variation curves drawn for the model are valid for the prototype and for each geometrically similar machine.

In the pages which follow, we shall use such adimensional coefficients and such curves to give the test results.

The variations of the torque and the power versus the rotational speed for different wind speeds can be deduced without difficulty from the preceding characteristics. From the previous relationships derive the following expressions:

$$F = \frac{1}{2} \rho C_F S V^2 \qquad\qquad P = \frac{1}{2} \rho C_p S V^3$$

$$M = \frac{1}{2} \rho C_m R S V^2 \qquad\qquad N = \frac{\lambda_0 V}{2 \pi R}$$

For a given machine, a given wind speed, and for every tip speed ratio λ_0, a value of C_F, C_m, C_p and therefore, a value of F, M and P can be found. Thus it is possible to draw the mechanical characteristics of the wind machine (thrust, torque and power) for different values of V versus the rotational speed N. These curves are essential to study the driving of pumps or electric generators by wind motors.

Remark concerning wind tunnel tests

The laws of similitude can be obtained by other methods, for example by expressing the geometrical similarity of the profiles, the kinematic similitude (same shape for the speed triangles) and dynamic similitude (the forces acting on the prototype and the model must be in the same ratio whatever may be their nature: inertia, viscosity, pressure, etc...).

Theoretically, it is necessary to take into account the viscosity forces. It amounts to the same thing, to express the equality of the Reynolds number on the prototype and the model.

$$\frac{V_1 D_1}{v_1} = \frac{V_2 D_2}{v_2} \quad \text{and} \quad \frac{U_1 D_1}{v_1} = \frac{U_2 D_2}{v_2}$$

One realizes that it is generally impossible to look into this condition for the big wind turbines. In fact, the model tests are generally made in wind tunnels at the atmospheric pressure and at the ambiant temperature. It results from this fact that $v_1 = v_2$, and therefore, the preceding equalities may be written as:

$$V_1 D_1 = V_2 D_2 \quad \text{and} \quad U_1 D_1 = U_2 D_2$$

As: $U_1 = \pi N_1 D_1$, and $U_2 = \pi N_2 D_2$, the last expression may still be written as: $N_1 D_1^2 = N_2 D_2^2$.

The first equation shows that the wind model must be tested in a wind having a speed $V_2 = V_1 \dfrac{D_1}{D_2}$ therefore higher than V_1. The second one points out that the rotational speed N_2 must be equal to: $N_2 = N_1 \dfrac{D_1^2}{D_2^2}$ therefore higher than N_1.

To realize the consequences of these assertions, take an example: Suppose we have made a model at the scale of 1/10 of a wind turbine having a diameter of 10 m. If the tip-speed ratio is $\lambda_0 = 6$, the rotational speed of the prototype is 100 r.p.m. in a wind having a speed of 8.70 m/s ($U_1 = 2\pi N_1 R_1 = \lambda_0 V_1$).

Thus to respect the Reynolds conditions, it would be necessary to test the model in a wind speed $V_2 = 87$ m/s ($V_1 D_1 = V_2 D_2$).

The rotational speed of the wind model would then be:

$$N_2 = N_1 (10)^2 = 100 \times 100 = 10\,000 \text{ r.p.m.}$$

At such a speed, compressibility phenomena would occur. The dynamic similitude would disappear because the phenomena produced on the model would be different from those which would appear in reality, on the full scale unit.

In practice, the model will be tested in a wind tunnel at wind speeds of the same level or slightly higher than those which the prototype would

experience in reality. Thus, the Reynolds numbers relative to the model
$(W_2 l_2/\nu)$ will be lower than those observed on the prototype $(W_1 l_1/\nu)$.
However, if the values of R remain higher than a certain value R_c called
Reynolds critical number, full similitude is attained. As a matter of fact,
above the value R_c, the drag coefficient C_d does not vary much with the
Reynolds number, so that the values of C_d for the model and for the prototype are practically equal for the same incidence angle. Consequently,
in this case, the above relationships are confirmed.

On the other hand, if the model tests are realized in such conditions
that R is lower than R_c, the drag coefficient C_d will be higher, at the same
incidence, on the model than on the prototype because of the greater influence of viscosity. The similitude will be less perfect.

It should be noted that the aerodynamic profiles of the modern wind
turbines are always chosen, for better efficiency (lowest values of the ratio
C_d/C_l), in such a manner that, when the wind turbine is running normally,
the Reynolds number in every section, is higher than the Reynolds critical
number R_c. This is about 10^4 for the curved thin profiles and varies
from 10^5 to 10^6 for the NACA profiles.

(a) Rotating-roofed windmill

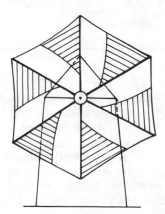

(b) Rotating-caged windmill

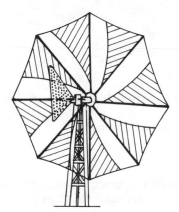

(c) Cretan wind turbine

(d) Portuguese windmill

Fig. 44

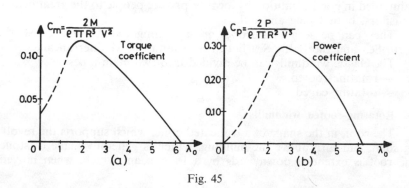

$$C_m = \frac{2M}{\rho \pi R^3 V^2}$$

Torque coefficient

$$C_p = \frac{2P}{\rho \pi R^2 V^3}$$

Power coefficient

(a)

(b)

Fig. 45

CHAPTER III

DESCRIPTION AND PERFORMANCES
OF THE HORIZONTAL-AXIS
WIND MACHINES

Most powerful windmill stations which have been built are based on the horizontal-axis system.

In this type of machine, one can differentiate among :
— the classic windmills,
— the slow wind turbines,
— the fast wind turbines.

In this chapter, we shall study the different machines sequentially, then we shall describe the direction and speed devices used in practice.

1. THE CLASSIC HORIZONTAL-AXIS WINDMILL

Many have disappeared but several still remain and are religiously maintained in good condition by local or private people to the great delight of visitors, both young and old.

They can be seen in Europe, mainly along the Atlantic coast, the North Sea and the Baltic Sea but also around the Mediterranean.

The classic windmills can be divided into two main types :
— rotating-roofed,
— rotating-caged.

Rotating-roofed windmills :

The roof, in the shape of a truncated cone, which supports the revolving shaft, can turn above the building, which is generally made of stone. The roof is extended downwards by a long beam which, when moved, allows the sails of the mill to be oriented to the wind.   The newest ones are directed by an auxiliary wind wheel.

Rotating-caged windmills :

The shaft of the wheel is linked to the cage which contains the mill stones.   The whole mill is placed on a spindle.   As in the previous type, the orientation of the wheel towards the wind is brought about by a direction arm or an auxiliary wind wheel.

When the mills were still working, the sails, usually made of wood, were covered with linen which was rapidly furled if a sudden gale arose. In some others, the rotational speed was adjusted by opening, to a greater or lesser degree, mobile shutters set in the sails.

In some Portuguese mills, the sails were made of triangular jib-shaped canvas, stretched between 8 or 10 poles arranged radially and supported by cross-pieces fixed on the shaft.

Generally, the windmill could be stopped from outside by a rope working on the brake which tightened against a cylinder inside the mill.

Characteristics

The sails were usually between 5 and 15 m long.   Their width was about one-fifth of their length.   Their rotational speed ranged between 10 and 40 r.p.m., the lowest speeds corresponding to the longest sails.

Tests done at the Eiffel Laboratory, in Paris, on a scale model demonstrated that the efficiency of the horizontal axis windmill was optimal with tip-speed ratio $\lambda_0 = U_0/V$ varying from 2 to 3.

With the tested machine (see graph 45b), the efficiency reaches a maximum when :

$$\lambda_0 = \frac{\pi DN}{60\,V} = 2.7$$

This corresponds to a rotational speed: $N = 51.5\ V/D$ and to a coefficient C_p equal to 0.3, i.e. to an effective amount of energy equal to 50 % of Betz' limit (C_p Betz $= 16/27 = 0.595$).

Under these conditions, the maximum power produced is given, as a function of the diameter and the wind speed, by the relation :

$$P = 0.15\ D^2 V^3$$

P being expressed in watts, D and V in meters and meter/s, ρ being equal to about 1.27 kg/m^3.

The application of the previous laws to windmills which have sails from 5 to 15 m long, leads to the figures collected in table 7, assuming a 7 m/s wind.

TABLE 7

Diameter in m	Wind speed in m/s	Max power in kW	Rotational speed in r.p.m
10 m	7	5.1	36
15	7	11.6	24
20	7	20.4	18
30	7	47	12

Note that only the descending part of the variation curve of the coefficient C_m corresponds to a stable operation of the machine.

2. SLOW WIND TURBINES

Since 1870, the multi-bladed low-speed wind turbine has appeared, first in America and then in Europe.

The blades, which vary in number from 12 to 24, cover the whole surface of the wheel, or almost. The tail vane, behind the windmill, keeps the wheel facing the wind. Figure 46 represents this type of windmill.

Features:

The diameters of the biggest windmills of this type usually built range from 5 to 8 m. A multi-bladed 15 m diameter windmill was even built in the United States. These multi-bladed windmills are particularly well adapted to low wind velocities. They start freely with winds ranging from 2 to 3 m/s. The starting torque is relatively high.

The variation graphs 47a and 47b represent the results of tests carried out at the Eiffel Laboratory in Paris.

In the case of the model studied, the production of energy is maximal when $\lambda_0 = 1$. These conditions correspond to an optimal rotational speed in revolutions per minute equal to:

$$N = \frac{60\,V}{\pi D} \simeq 19\,\frac{V}{D}$$ and a C_p equal to 0.3, i.e. an effective amount of energy equal to 50 % of the Betz limit.

When taking $\rho = 1.27$ kg/m^3 as the value of the air's specific mass, it follows that the maximum power likely to be produced by this type of machine can be calculated in relation to the diameter by an expression similar to that which gives the power of windmills:

$$P = 0.15\,D^2\,V^3$$

the power being expressed in watts, the diameter in m, and the wind speed in m/s.

By applying the above relation to machines of different diameters and by considering 5 to 7 m/s winds, we obtain the speed and the power values indicated in table 8.

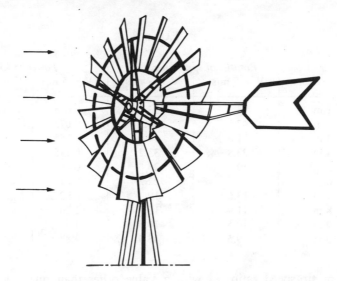

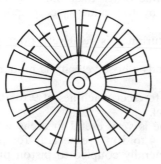

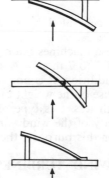

Various methods of blade fixation

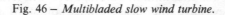

Fig. 46 – *Multibladed slow wind turbine.*

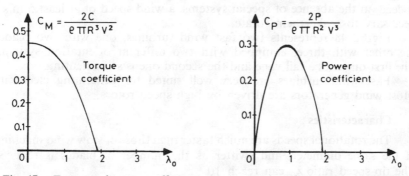

Fig. 47 – *Torque and power coefficients of a slow wind machine as a function of λ_0.*

TABLE 8

Diameter of the wind wheel in m	Rotational speed in r.p.m		Power in kW	
	V = 5 m/s	V = 7 m/s	V = 5 m/s	V = 7 m/s
1 m	95	133	0.018	0.05
2 m	47.5	66.5	0.073	0.40
3 m	31.9	44.5	0.165	0.45
4 m	23.8	33.2	0.295	0.81
5 m	19	26.6	0.46	1.26
6 m	16	22.2	0.67	1.8
7 m	13.6	19	0.92	2.5
8 m	11.9	16.6	1.20	3.3
9 m	10.5	14.8	1.52	4.2
10 m	9.5	13.3	1.87	5.15

For tip-speed ratios λ_0 with a value other than one, the indicated rotational speeds must be multiplied by the value λ_0 corresponding to the machine.

Power generated by slow wind turbines is relatively poor for two reasons :

— These machines mainly use winds whose velocity is moderate and varies from 3 to 7 m/s.

— In addition, it is not very easy to erect machines of 9 to 10 m diameter because of the weight of the wheel.

Nevertheless, this type of machine is very useful in areas where the mean velocity of the wind varies from 4 to 5 m/s, especially for pumping water. For this purpose, they are generally coupled to piston pumps.

3. FAST WIND TURBINES

In this type of wind machine, the number of blades is much more limited, varying from 2 to 4. At equal power, these windmills are much lighter than the slower ones and, therefore, more interesting for that reason.

However, they have the disadvantage of starting with difficulty. Indeed, in the absence of special systems, a wind speed of at least 5 m/s is necessary to make them rotate.

Figure 48 represents two fast wind turbines, one with two blades, the other with three, equipped with two different orientation systems : The first one uses a tail vane and the second one is self-orienting.

Fast wind turbines are very well suited for generating electricity. Most wind generators are driven by high speed rotors.

Characteristics

The rotational speeds are much faster than those of slow wind machines of the same diameter, and swifter as the number of blades is reduced. The tip-speed ratio λ_0 can reach 10.

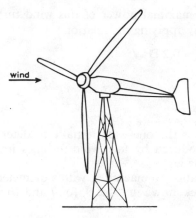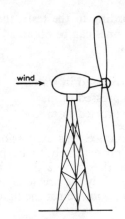

Three-bladed wind rotor   *Self-orienting two-*
with a tail vane.   *bladed wind rotor.*

Fig. 48

At equal wind speed and equal diameter, the torque developed by a high-speed wind rotor is less than that provided by a slow wind machine.

Figures 49a and 49b exhibit the variation curves and power coefficients in relation to the tip-speed ratio λ_0 of a two-bladed wind rotor tested at the Eiffel laboratory in Paris.

The wind turbine has an optimal output when:

$$\lambda_0 = \frac{\pi DN}{60 V} = 6$$

This corresponds to a rotational speed $N = 115 \frac{V}{D}$ and to C_p equal to 0.4.

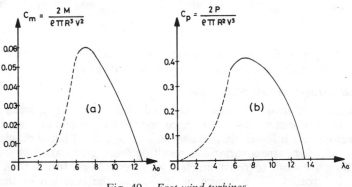

Fig. 49 – *Fast wind turbines.*
Torque and Power coefficients.

According to the tests, the maximal power of this wind turbine and of similar ones can be obtained by applying the relation :

$$P = 0.2 \, D^2 \, V^3$$

with P expressed in watts,
   D expressed in m,
   V expressed in m/s.

In practice, this expression is the one used initially to determine the maximal power likely to be produced by fast wind turbines irrespective of whether they have 2, 3 or 4 blades.

If we apply the previous relation to machines, with a diameter varying from 2 to 50 m, we get the figures shown in table 9, for 7 and 10 m/s wind speeds.

TABLE 9

Diameter in m	Rotational speed in r.p.m		Max. Power in kW	
	V = 7 m/s	V = 10 m/s	V = 7 m/s	V = 10 m/s
1	935	1 340	0.07	0.2
2	470	670	0.27	0.8
3	310	450	0.60	1.8
4	235	335	1.07	3.2
5	190	270	1.7	5
6	155	220	2.4	7.2
8	120	168	4.4	12.8
10	95	134	6.7	20
15	62	90	15	45
20	47	67	26.8	80
30	31	45	60	180
40	23	33	107	320
50	19	27	168	500

Note that a 12.6 m/s wind speed would produce a quantity of energy equal to twice the calculated values for V = 10 m/s.

Advantages of fast wind rotors

Because of their fast rotational speed, they have only a small number of blades : 2, 3 or at most 4. The price and the weight of a fast wind turbine is therefore much lower than those of a slow one of equal diameter.

In addition, stress variations resulting from gusts are less important since they are built to resist much higher centrifugal forces than are the slow wind rotors. The furling devices using the rotation of the blades on their axes, for use in storms, also require less energy.

Another advantage :

When the machine is kept motionless, the axial thrust (even if the blades are in working position) is lower than when it is running. This is not the case for slow windmills.

Readings taken in Denmark on the Gedser three-bladed windmill have proved that the axial force, when the machine is stationary, reaches 40 % of the force exerted on the turning windmill.

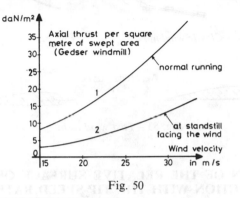

Fig. 50

Drawbacks

These advantages are counterbalanced by a drawback; the starting torque is low. A rapid wind rotor must start without too much effort. This drawback can be limited by giving the blades near the axis a sufficient chord and the best pitch angle possible. Variable pitched blades with a regulator such as the Aerowatt model for instance, can also be used. The pitch angle is maximum when the machine starts and declines as speed increases.

4. PROFILES USED

These change according to the type of machine :

The sails of the first windmills were made of timber-work covered with linen. The sails were supported by wooden bars on both sides of the stock. Later, the bars were moved to the trailing edge of the wings to improve the aerodynamic efficiency. More modern designs substituted sheet metal for the cloth sails and used shutters and flaps to control the speed of the rotor in high winds.

Slow windmills use thin and slightly concave profiles. Because of their low rigidity, these profiles are fixed to a metallic circular frame forming the skeleton of the mobile wheel.

Fast-running wind turbines always have streamlined blades. Their profiles are usually chosen from the NACA or Göttingen series (NACA 4412, 4415, 4418, 23012, 23015, 23018, Göttingen 623, 624). These aerofoils are characterized by reduced drag and also by a high aerodynamic efficiency provided that the Reynolds number is high enough.

In some recent high-speed wind power plants, laminar flow profiles of the Wortmann and NACA series are also used (FX 60.126, 61.140, 77..., NACA 63_2615, 63_2618, 64_2612, 64_2618). The drag/lift ratios of these aerofoils are very low. Therefore, the use of such profiles can lead to very good performances.

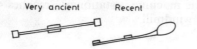

Fig. 51 – *Classic windmills.*   Fig. 52 – *Slow wind rotor*

Fig. 53 – *Fast wind turbine.*

5. VARIATION OF THE RELATIVE SURFACE OF THE BLADES IN CONNECTION WITH THE TIP-SPEED RATIO: λ_0

The ratio of the total surface of the blades to the area swept by the rotor decreases as the tip-speed ratio rises, as shown in graph 54. This graph is copied from a research paper presented by Ulrich Hütter at the New Delhi Congress in October 1954.

6. ORIENTATION SYSTEMS

The most frequently used are :
— the tail vane,
— the auxiliary wind wheel,
— the wind turbine downwind of the support,
— servo-motor systems,
— manual systems.

a) THE TAIL VANE

This device is mainly used to orient slow wind turbines of up to 6 m diameter.

To get a satisfactory result, certain conditions have to be met.

Let E equal the distance between the orientation axis (pivot) and the propeller rotation plane. If a value equal to 4E is given to L, the distance between the orientation axis and the centre of tail vane, the surface A of the tail vane must have the following values in relation to the area S swept by the rotor :

For a multi-bladed windmill, A = 0.10 S.
For a two or three bladed windmill, A = 0.04 S.

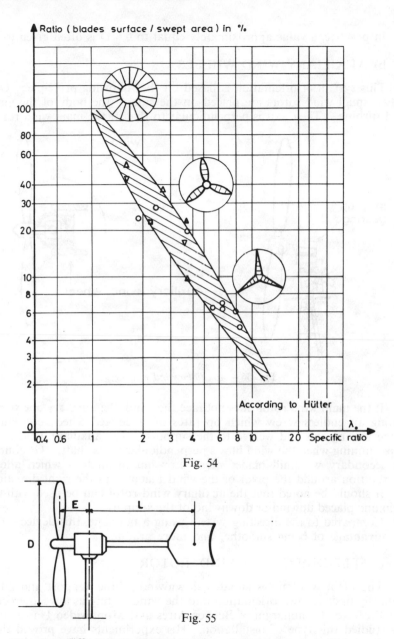

Fig. 54

Fig. 55

When L differs from 4E, the area of the rudder needed to ensure stability can be calculated from the following equations:

$$S = 0.40 \ S \ \frac{E}{L} \text{ for a multi-bladed wind rotor,}$$

$$S = 0.16 \ S \ \frac{E}{L} \text{ for a fast one.}$$

In practice, a value approximately equal to 0.6 D is often given to L.

b) AUXILIARY WIND WHEELS

This system of orientation is based on the following principle : One or two small wind rotors are placed on the side of the body of the main wind turbine. Their axis is perpendicular to that of the main wind rotor.

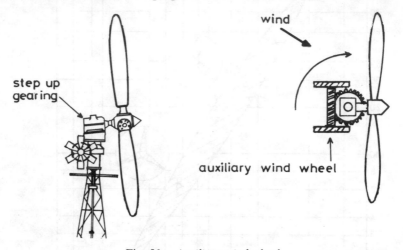

Fig. 56 – *Auxiliary wind wheel.*

If the main wind rotor does not face the wind, the auxiliary one starts, driving an endless screw which operates on a cogged wheel, concentric to the yawing axis and welded to the support. The auxiliary wind rotor stops running when the wind blows perpendicular to its shaft. Of course, the secondary windmill blades must show an inclination which allows the rotation around the pivot of the wind machine in the right direction.

It should be noted that the auxiliary wind rotor can be used with the main one placed upwind or downwind of the support.

Compared to the directing system using a tail vane, this device offers the advantage of being smoother and more gradual.

c) SELF-DIRECTING WIND ROTOR

The main windmill is situated downwind of the support and automatically finds its own orientation into the wind, acting as a weathercock.

Professor Cambilargiu of Buenos Aires and Montevideo Universities has studied this type of installation. His experiments have proved that, taking a support with a diameter $d = 0.022$ D (D being the diameter of the windmill), vibrations are noticed when the propeller plane is situated at a distance x lower than 0.25-0.30 D from the pivot axis. For a streamlined support with a relative span equal to 3, and for a two-bladed rotor, this distance is reduced to 0.13 D. In Professor Cambilargiu's opinion, the vibrations would be even further reduced with a three-bladed rotor.

d) ORIENTATION BY SERVO-MOTOR

The control arrangements of such a system, which uses a servo-motor, can be made as shown in fig. 57.

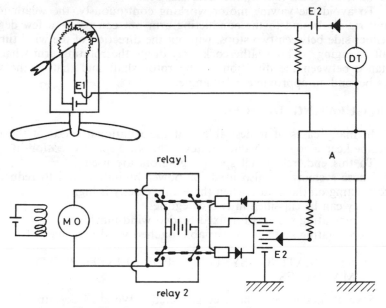

Fig. 57 – *Device using an electrical servomotor.*

The yawing motor, which can turn both ways, is controlled by a weathercock and by a tachometrical dynamo driven by the windmill. The weathercock, fixed on the frame of the windmill above the wind rotor, carries a conducting arm which moves around a spindle on a horizontal circular rheostat forming one piece with the windmill frame. The tension between the middle point of this rheostat and the conducting arm is applied via a resistance R on an amplifier.

The tachometrical dynamo DT, which produces a tension proportional to the rotational speed, feeds a circuit which includes the previous resistance R, a rectifier and a battery E2 placed in opposition.

As long as the rotor speed is lower than the nominal speed, the tension between the tachometrical dynamo poles is weaker than the voltage between the battery E2 poles. The dynamo delivers no current. The tension introduced into the amplifier entrance is then the one which exists between M and P. The yawing motor MO starts and operates until the windmill takes up its position right into the wind.

If the rotational speed is too high, the dynamo delivers a current which causes a dropping of the tension inside R. The yawing motor starts at that moment. The wind rotor moves aside from the wind direction until the sum of the voltages at the poles of R and MP reaches zero.

The rotational speed is then limited proportionally to the reduction of the surface facing the wind.

This device prevents the wind rotor from racing when a suppression or a reduction of the load on the rotor axis occurs.

To avoid the yawing motor working continuously, the weathercock is not rigidly bound to the conducting arm but can rotate a few degrees to either side between two stops, without the direction of the wind turbine shaft changing. The weathercock only drags the rotating arm when the distance between the direction of the rotor shaft and that of the wind goes beyond an approximate 10° angle.

7. REGULATING DEVICES

In many cases, it is desirable that the rotational speed of the wind machine keep a constant value whatever the wind speed variations may be.

To this end, rotational speed regulators are used.

These systems are also used for power limitation and to reduce the forces acting on the blades when the wind velocity is high.

They can be classified in two categories :
— regulating systems for fixed-bladed wind machines,
— regulating systems for variable-pitched wind machines

a) REGULATING SYSTEMS FOR FIXED−BLADE WIND MACHINES

Several models have been constructed. We shall describe the main ones :
— Devices using an articulated tail vane and rotating the turbine out of the wind around the vertical axis.
— Devices using a rigid tail vane and rotating the turbine out of the wind around a horizontal axis.
— Devices using a rigid tail vane and an aerodynamic brake.

1°) Devices using an articulated tail vane and turning the turbine shaft out of the wind direction by rotation around the vertical axis

To avoid the racing of the rotor as the wind velocity increases, the air flow intercepted by the rotor is reduced by rotating the turbine shaft out of the wind direction. If S is the swept area and θ the angle between the wind direction and the axis of the machine, then the intercepted area in the airflow is S cos θ.

If the wind velocity becomes excessive, the wind machine shaft rotates by 90° about the vertical axis and the wind rotor stops.

Two types of mechanisms are used to obtain the desired effect. They are shown in fig. 58 and 59.

First device

The tail vane is articulated and joined to the body of the machine by a spring. The aerodynamic force acting on a lateral plane fastened rigidly to the rotor body moves the wind turbine axis away from the wind direction. The machine takes a position such that the sum of the moments of the forces acting on the lateral plane, on the tail vane and on the wind rotor equals zero.

To reduce oscillations, a shock absorber is sometimes set on the tail vane joint.

This system is generally used in the regulation of water pumping wind machines.

Second device

In this system, there is no lateral plane. The rotor is mounted eccentrically to the orientation axis (yawing axis). The spilling moment is due to the aerodynamic forces acting on the rotor. As a consequence of the rotor shifting, the moment of aerodynamic forces about the yawing axis is different from zero.

The inclination of the rotor axis relative to the wind direction diminishes the intercepted wind flow. Thus the output is reduced.

It must be noted that this second device is the only one used in practice on slow wind machines of large diameter.

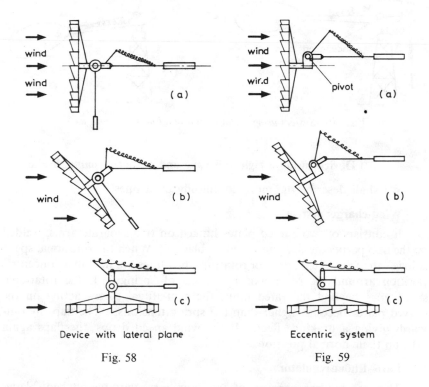

Device with lateral plane   Eccentric system

Fig. 58   Fig. 59

Regulating systems using an articulated rudder.

2°) **Device using a rigid tail vane and turning the turbine out of the wind by rotation around a horizontal axis**

This system is shown in fig. 60.   The turbine can turn by 90° around a horizontal axis perpendicular to the turbine shaft.

If the wind velocity is moderate, the machine shaft remains horizontal, but, if the wind speed is high, the rotor shaft comes out of the wind direction by rotating around the horizontal axis.   The wind machine takes a position such that the sum of the moments due to the aerodynamic forces, to the spring tension and to the rotor weight is equal to zero.   To avoid oscillations, it is desirable to add a damper.

This device is only used on aerogenerators up to 4 m in diameter.

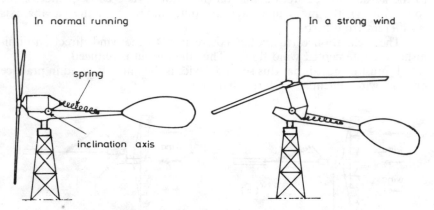

Fig. 60 – *System using inclination around an horizontal axis.*

3°) **Devices using a rigid tail vane and an aerodynamic brake**

We shall describe the more commonly used ones :

Wind-charger system

It consists of two curved plates hinged on two separate arms, welded on the hub perpendicularly to the main blades.   When the rotational speed is lower than the rated speed of rotation, the plates are held in a concentric position around the rotor axis by means of springs.   If the rotational speed goes beyond the rated limit, the centrifugal forces acting on the curved plates become higher than the spring tensions.   The flaps fly outwards giving a braking effect.   If the wind speed drops, the flaps again take up their normal position.

Paris-Rhône regulator

This mechanism consists of two auxiliary variable-pitched blades fastened on the hub and perpendicular to the main ones.   These auxiliary blades provide a high starting torque because of their high pitch angle

Moderate wind   Strong wind

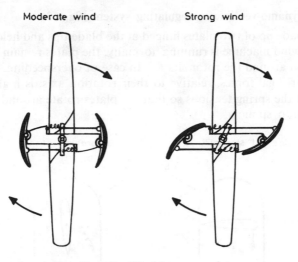

Fig. 61 – *The Windcharger regulator.*

at rest.   As the rotor accelerates, they are subjected to centrifugal forces
and they turn around their own axis because each of them has a helical
socket and a spring at its root.

In normal running, they do not supply an important contribution to
the output power.   But in case of overspeeding, under the increasing
centrifugal forces which act on them, their pitch angles decrease to zero and
become negative.   They turn into an aerodynamic brake and then reduce
the rotational speed.

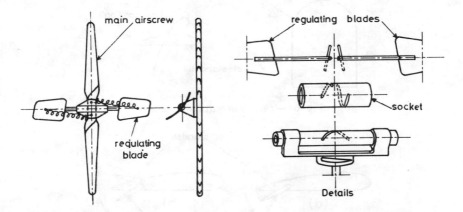

Fig. 62 – *The Paris Rhône regulator.*

Aerodynamo-ventimotor regulating system (fig. 63)

It is made up of two plates hinged at the blade tips and held by springs. When the wind machine is running normally, the plates remain in a concentric position around the rotor shaft. In case of overspeeding, the moment of the centrifugal forces, relative to their rotation axis, is higher than the moment of the spring tensions so that the plates rotate around their hinges and turn into spoilers.

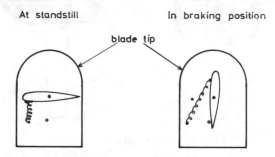

Fig. 63 – *Ventimotor aerodynamic brake.*

Danish regulating system (fig. 64)

The end of the blade, whose area is about one-tenth of the whole blade surface, can turn around a radial axis. In normal running, each blade tip is in alignment with the other part of the blade and provides an important contribution to the output torque. In case of overspeeding, or to stop the machine, the end of the blade can rotate up to 60° or 90° around the

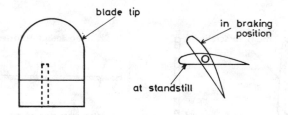

Fig. 64 – *Danish aerodynamic brake.*

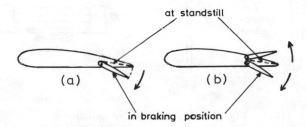

Fig. 65 – *Regulating system using flettners.*

control axis, and then produces a braking effect. The rotation of the blade tips is achieved by means of a servomotor or by a helical socket and a spring.

Another system based on the same principle uses flettners.

Hollow-bladed device

The blades are hollow and the inner channels are connected to the hub as shown in fig. 66. In normal running the channels are shut by a valve placed on the rotor hub or by several valves placed at the blade tips. The air cannot pass inside the blades.

In case of overspeeding, the valves open under the action of centrifugal forces acting on the device and the air subjected to centrifugal force flows in the channel from the hub towards the blade tips.

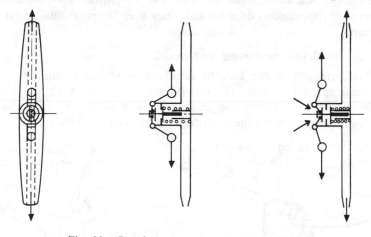

Fig. 66 – *Regulating system using hollow blades.*

b) REGULATING SYSTEMS FOR VARIABLE-PITCHED WIND MACHINES

The machines with variable-pitched blades allow a greater flexibility of operation.

A wide range of pitch-control mechanisms is currently in use. Some of them use springs, weights, flettners and servomotors. The variations in the blade pitch are controlled either by variations of aerodynamic pressure or by centrifugal forces acting on weights whirling with the rotor. Centrifugal regulators are more usual. Among the firms using them, we can mention Quirk, Enag, Aerowatt, Elektro and Jacobs.

1°) **Pitch-control device based on the variation of the aerodynamic thrust on the blades** (fig. 67)

The regulator which uses a spring system balancing the aerodynamic thrust arising on the blades has the advantage of behaving very well in gusts. With this system, the position of the pitch-change axis is chosen so that the aerodynamic forces acting on the blade, create a torque which is always in the same direction. This torque is balanced by the action of a spring so that the rotational speed is kept approximately constant.

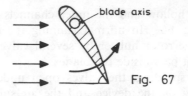

Fig. 67

However, the centrifugal devices whose description follows, are often preferred to the previous ones because they keep the rotational speed more constant.

2°) **Elektro regulating system**

The principle of the Elektro regulator is shown in fig. 68. If the rotational speed increases, the centrifugal force acting on the rotating weight compresses a spring and makes the pitch blade vary. The balance position is obtained when the centrifugal force is equal to the spring tension.

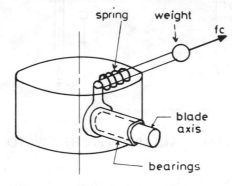

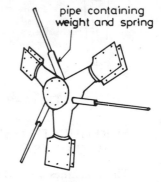

Fig. 68 – *Principle.*   Fig. 69 – *Overall view*

The Elektro regulator.

The Elektro regulator shown in fig. 69 is based on this system. Between the cut-in-speed and the rated rotational speed, the pitch angle remains constant. If the rotational speed exceeds a certain limit, the pitch angle increases to maintain the rotational speed at a constant value.

3°) **Enag and Quirk regulator**

In case of overspeeding, under the action of centrifugal forces, the regulation weights fastened on each blade move away from the rotation axis, making the blades turn around the pitch-change axis. The pitch angles increase and the rotational speed decreases, coming back approximately to its previous value. The balance is obtained when the centrifugal moment is equal to the opposite moment of the spring force.

This device, which uses a single spring, is particularly reliable and sturdy.

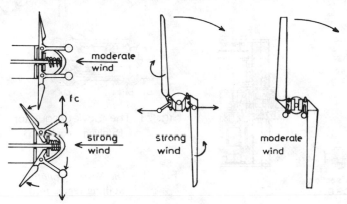

Fig. 70 – *Enag and Quirk regulators Principle of operation*
(Journal « Écologie »)

4°) **Stalling regulating system:** Soviet regulator and Aerowatt system (fig. 71 and 72).

The devices used by the Soviets as well as by the Aerowatt French Company are based on the same principles. All of them include regulation weights and two springs: the first fairly supple and the second much more rigid.

Under the action of the centrifugal forces exerted on the weights, the springs are compressed and the pitch decreases as indicated in figure 71.

At rest, pitch is high. Thus the machine starts without difficulty at wind speeds over 3 m/s.

As the rotational speed increases, the more supple spring is compressed, while the length of the second one remains unaltered, and the pitch angle decreases. It stays at a constant value when the first spring is completely compressed provided that the number of revolutions per minute does not exceed the rated rotational speed.

In case of rotor overspeeding, the second spring is also compressed. The pitch angle at the blade tips reaches zero and can take negative values. The incidence angles increases behind the stalling point on the Eiffel polar which corresponds to a dropping of the lift force. The output torque is reduced and the rotational speed drops. The blade tips act as an aerodynamic brake.

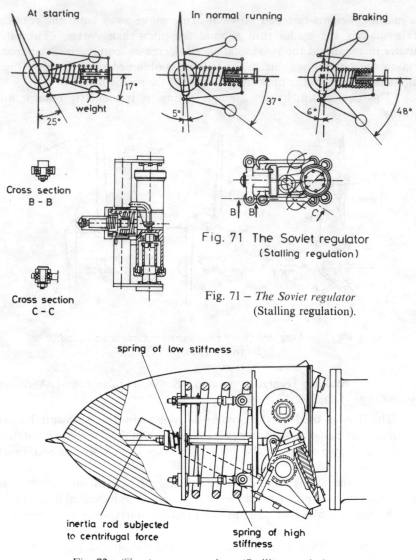

At starting

In normal running

Braking

weight

17°

25°

5°

37°

6°

48°

Cross section
B - B

Cross section
C - C

Fig. 71 The Soviet regulator
(Stalling regulation)

B B

C

Fig. 71 — *The Soviet regulator*
(Stalling regulation).

spring of low stiffness

inertia rod subjected
to centrifugal force

spring of high
stiffness

Fig. 72 — *The Aerowatt regulator* (Stalling regulation).

5°) **Fly-ball regulator** (fig. 73)

The pitch may be changed by means of a fly-ball governor. This centrifugal spring-controlled device is used on the German Allgaïer wind machines and on the Soviet Sokol wind machine.

The variation of the pitch angle is controlled through the shaft of the machine which is hollow.

The pitch remains unaltered until the machine has reached its rated rotational speed.

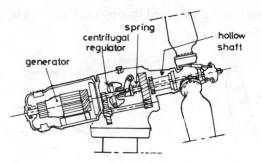

Fig. 73 – *Head of the Allgaier wind machine.*

Beyond the speed limit, the fly-ball governor enlarges the pitch, and the rotational speed remains approximately constant.

6º) Jacobs regulator

In this regulator, the pitch variation is controlled by centrifugal forces acting on masses, through pinions which gear in a cogged wheel, in the hub.

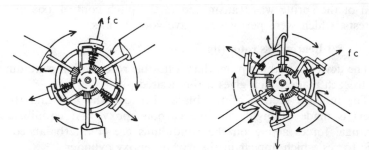

Fig. 74 – *The Jacobs regulator.*

When the rotational speed increases, the centrifugal forces, which are counterbalanced by spring tensions, tend towards an increase in the pitch. Therefore the rotational speed remains moderate. This system, which is used for regulation of three-bladed wind machines, is effective up to 40 m/s wind speed.

7º) Bent-bladed wind machines with variable pitch

This arrangement was described by A. Lacroix in the journal : "La Technique Moderne" of August 1949. The blade axis is bent so that the centre of inertia is not placed on the pitch-change axis.

As the rotational speed increases, the centrifugal forces which act on the blades make the pitch vary. The aerodynamic force arising on the blades also acts in the same way but with lesser effect. The regulation of the rotational speed is obtained by opposing a spring to the action of the foregoing forces.

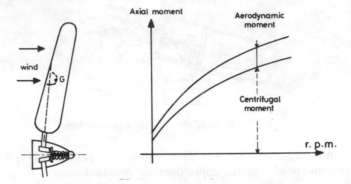

Fig. 75 – *Device using blades as a governor.*

This device was tested in Japan by Moriya and Tomasawo. The action of the spring was transmitted to the pitch control axis by a cam. The tests, which were performed, gave good results.

8°) **Bearingless rotor** (fig. 75)

The device is nothing more than a flexible rotor with a pendulum attached to each blade which gives it the desired pitch.

The inboard section of the blade (between the hub and the blade proper) is made of highly flexible carbon epoxy. At equilibrium, the centrifugal forces acting on the pendulums are counterbalanced by the elastic forces which appear in the carbon-epoxy cylinder.

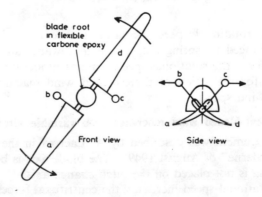

Fig. 76 – *Bearingless rotor.*

9°) Soviet stabilizer regulation system

This system, conceived by Sabinin and Krabowsky, was tested on wind machines of 24 and 30 m diameter.

The aerodynamic force acting on a streamlined element called the stabilizer is used to rotate the blade tip or the whole blade around its pitch change axis. The stabilizer plays the role of an aerodynamic servo-motor. The angle of rotation of the blade is determined by the stabilizer whose position is controlled by a centrifugal regulator.

At rest, the blades are feathered. When the spring of the regulator is released, the stabilizer takes an inclined position relative to the wind direction which forces the blade tip to rotate around the pitch-change axis. If the wind speed is high enough, the rotor starts and the pitch diminishes.

If the rotational speed increases too much, the stabilizer acts in the opposite sense by increasing the pitch and thus reducing the rotational speed. Weights fastened on the blades diminish the effort required for regulation.

The system has a high sensitivity and provides a fairly constant rotational speed.

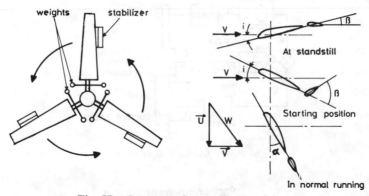

Fig. 77 – *Soviet regulator using a stabilizer.*

10°) Electronic devices

Figures 78a, b, c represent some theoretical diagrams of modern electronic regulating systems indicated by I. K. Buehring and L. L. Freris, from the University of Exeter and Imperial College (U.K.) respectively.

They are based on the control of different parameters. The first device has been conceived to hold the rotational speed at a constant value. The second controls the output power, and the third keeps the tip speed ratio λ_0 constant. The regulation is made by varying the firing angle of a thyristor rectifier.

The previous devices may be used for regulating wind power plants running autonomously. They must be completed with a system limiting power output, for example, with a system producing a change in pitch or a braking action for a wind rotor with fixed blades.

Note that nowadays, the tendency is to control wind-driven generators by means of microprocessors.

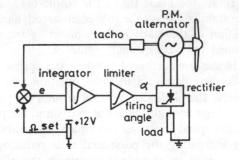

(a) *Constant speed controller*

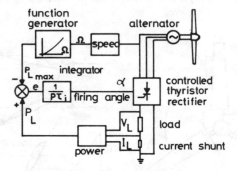

(b) *Power feedback controller.*

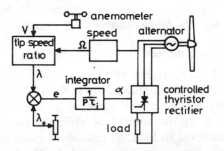

(c) *Tip-speed ratio controller.*

Fig. 78 – *Electronic devices.*

8. OTHER MODELS OF HORIZONTAL-AXIS WIND MACHINES

The above wind machines are the most commonly used. However, others models working with a horizontal axis have been built, but some of them are only prototypes. There is the single-bladed machine, the horizontal-axis wind turbine with other wind rotors fastened to its blades the diffuser augmenter, the dynamic inducer and the tornado system.

Single-bladed wind machine

This model has the advantages of permitting simple regulation and being cheap, but its shape is unaesthetic. In addition, the bending moment at the blade root is considerable and the vibrations are more severe.

Horizontal-axis wind turbine having other rotors on its blades

This arrangement has the advantage of suppressing the gearing between the usual wind rotor and the generator, but the proposed design has not been very successful so far.

Diffuser augmenter

Experiments have been undertaken on shrouded wind machines. They showed that the most effective design was constituted by the diffuser, which is placed just behind the wind rotor as shown in fig. 80. The diffuser is narrow at the point where it faces into the wind and expands as it goes back.

This configuration creates an important pressure drop behind the rotor blades causing increased airflow through the propeller. Tests have shown that the wind speed through the rotor was multiplied by 1.5 and the power by 3.5.

In fact, if we consider the output power produced by a windmill having a diameter equal to the output diameter of the diffuser, we observe that the power provided by this is approximately equal to the power supplied by the diffuser augmenter system.

Dynamic inducer

The dynamic inducer consists of a small T-shaped device appended to the tips of the rotor blades of a conventional horizontal axis machine. This configuration pushes the air out away from the propeller, thereby causing more air to be drawn into the propeller and yielding higher efficiency. To date, the concept has not known great success.

Tornado system

Dr. Yen's system consists of a tower with openable vertical vents. The vents on the side towards the wind are open while those opposite to the flow are closed.

As the wind blows into the tower, it rotates towards the top creating

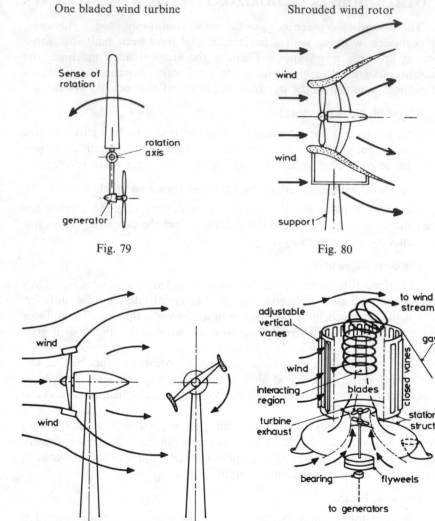

One bladed wind turbine

Shrouded wind rotor

Fig. 79

Fig. 80

Fig. 81 – *Dynamic inducer.*

Fig. 82 – *The tornado system.*

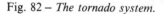

a small tornado.   In the centre of the vortex, the pressure is low, so outside air is sucked through the openings around the base of the tower and drives a wind rotor placed inside a shroud.

Dr. Yen thinks that the power provided by the wind rotor will be equal to a hundred times the power delivered by the same rotor placed in an isolated position.

9. BLADE CONSTRUCTION

The material used for blade construction must be strong and light and must not be subject to deterioration in bad climatic conditions. The strength to weight ratio must have the highest possible value.

In practice, blades can be made of wood, aluminium, fibreglass-reinforced plastic or steel.

Wood is used to construct the blades of small wind generators. The blades are carved and fastened onto the hub with steel bolts. For the blades of medium-size machines, wood may be used in the form of plywood skins in the form of sheets of laminated wood coated with resorcinol adhesive. Wooden blades must be absolutely waterproof because moisture makes imbalance possible and deteriorates wood quality. To protect against moisture, wood can be coated with fibreglass and resin or covered with varnish.

Aluminium alloy in the form of extruded bars is also commonly used in the construction of small wind machines.

To make blades of medium and large wind rotors, fibreglass is generally favoured by manufacturers. It is weather-resistant and can be easily shaped.

Composite blades made of steel and fibreglass-reinforced plastic are also manufactured.

Aluminium alloy and stainless steel are the most likely alternatives for very large wind rotors, because the blades made of these materials are more rigid than those consisting of fibreglass-reinforced plastic.

For the blades of slow-wind machines, galvanized iron sheet is generally employed.

CHAPTER IV

HORIZONTAL-AXIS WIND TURBINES
DESIGN OF THE BLADES AND
DETERMINATION OF THE FORCES
ACTING ON THE WIND POWER PLANT

The main element of a wind machine is the rotor.   Before undertaking its construction, two problems have to be solved :
— first the definition of the aerodynamic configuration of the blades, choice of profile, chord measurements, setting angles, number of blades, etc.
— next, the design of the blades with regard to material strength to permit the rotor to withstand the most severe operating conditions with an appropriate margin of safety.

A. Aerodynamic Configuration of the Rotor

NUMBER OF BLADES AND ROTOR DIAMETER

The principal layout of a machine is dictated by its purpose and the local wind velocity.

Thus if we intend to construct a wind-driven generator, to be erected on a windy site, it is better to choose a fast-running wind rotor having a high tip-speed ratio λ_0, say 5 to 8, in order to limit the gearing ratio. The wind machine will therefore be two or three-bladed.

If it is necessary to pump water in a country where the wind velocity is low or moderate, a multibladed wind turbine, with a tip-speed ratio varying from 1 to 2, is well suited.

In practice, the number of blades depends on the ratio λ_0 between the blade tip-speed and wind velocity upstream of the machine.

For $\lambda_0 = 1$ we shall adopt   8 to 24 blades
$\lambda_0 = 2$ 6 to 12
$\lambda_0 = 3$ 3 to 6
$\lambda_0 = 4$ 2 to 4
$\lambda_0 > 5$ 2 to 3

The diameter of the rotor will then be determined by one of the relations that we have seen in the preceding chapter :

$P = 0.15 \; D^2 V^3$ for a slow-running wind rotor
$P = 0.20 \; D^2 V^3$ for a fast-running wind turbine.

But finding the answer to these questions is not enough to enable us to build the wind rotor. For the construction, it is necessary also to know the blade chord and the setting angle at variable distance from the rotation axis.

In order to solve this problem, several theories have been established. We shall refer to the theories developped by Sabinin, Stefaniak, Hütter and Glauert. All of them introduce a system of vortices.

For our part, we shall limit ourselves to a simple theory and to the vortex theory of Glauert improved by the researchers of the University of Amherst, Massachussetts, USA.

In conclusion, we shall carry out a comparison between the results obtained by applying the different theories mentioned above. We shall see that all of them lead to very similar values.

I – THE SIMPLIFIED WINDMILL THEORY

1. DETERMINATION OF A BASIC RELATIONSHIP FOR CALCULATING THE BLADE CHORD

To determine the blade chord, we shall evaluate, the axial thrust on the section situated within the interval r, r + dr from the rotation axis, in two ways. This calculation will be carried out assuming the windmill is running in optimal conditions according to the Betz formula.

a) FIRST EVALUATION

According to the Betz theory, the axial thrust on the whole rotor is given by :

$$F = \frac{\rho S}{2} (V_1^2 - V_2^2)$$

and the wind speed through the rotor by:

$$V = \frac{V_1 + V_2}{2}$$

where V_1 and V_2, are the wind velocities at a certain distance in front of and behind the wind machine.

The power output reaches its maximum when $V_2 = \dfrac{V_1}{3}$.

The axial thrust F and the wind velocity V through the swept area, are then:

$$F = \frac{4}{9}\,\rho S V_1^2 = \rho S V^2 \quad \text{and} \quad V = \frac{2}{3}\,V_1$$

Assume that each element of the swept area affects the total axial thrust proportionally to its proper area. The contribution of the elements situated within the interval r, r + dr is:

$$dF = \rho V^2\, dS = 2\pi\rho V^2\, r dr.$$

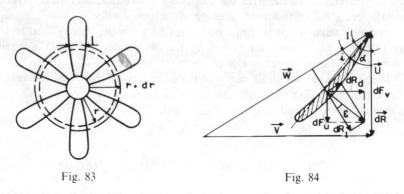

Fig. 83   Fig. 84

b) SECOND EVALUATION

The rotational speed being ω, the circumferential speed of the blade element at distance r, is: $U = \omega r$. The relationship between the absolute wind velocity $\vec{V}$ through the rotor, the circumferential speed $\vec{U}$ of the element at radius r and the air velocity $\vec{W}$ relative to the blade section: $\vec{V} = \vec{W} + \vec{U}$ can be written as: $\vec{W} = \vec{V} - \vec{U}$.

Estimate the aerodynamic forces which act on the element whose length is dr. We get for the lift and the drag forces:

$$dR_l = \frac{1}{2}\,\rho C_l W^2\, l dr$$

$$dR_d = \frac{1}{2}\,\rho C_d W^2\, l dr$$

and for the resultant:
$$dR = \frac{dR_l}{\cos \varepsilon}$$

ε being the angle between $\overrightarrow{dR}$ and $\overrightarrow{dR}_l$, l, the chord of the blade at distance r from the rotor axis.

As $W = \frac{V}{\sin I}$, it follows that:

$$dR = \frac{1}{2} \rho C_l \frac{W^2}{\cos \varepsilon} ldr = \frac{1}{2} \rho C_l \frac{V^2}{\sin^2 I} \frac{ldr}{\cos \varepsilon}$$

Let us project dR upon the rotor axis and calculate the contribution dF of the parts of the blades inside the interval r, r + dr in the total axial thrust.
b, being the number of blades, we get:

$$dF = \frac{1}{2} \rho C_l b \frac{V^2}{\sin^2 I} \frac{\cos (I - \varepsilon)}{\cos \varepsilon} ldr$$

On identifying this equation with the expression obtained for dF in the previous paragraph, it follows that:

$$C_l bl = 4\pi r \frac{\sin^2 I \cos \varepsilon}{\cos (I - \varepsilon)}$$

2. TRANSFORMATION OF THE PREVIOUS RELATIONSHIP. SIMPLIFICATION

On developing $\cos (I - \varepsilon)$, the preceding relationship can also be written as:

$$C_l bl = 4\pi r \frac{\tan^2 I \cos I}{1 + \tan \varepsilon \tan I}$$

In the optimal running conditions, the wind velocity through the rotor is $V = \frac{2}{3} V_1$.

Thus, the inclination angle is given by the equation:

$$\cot I = \frac{\omega r}{V} = \frac{3}{2} \frac{\omega r}{V_1} = \frac{3}{2} \lambda.$$

By carrying back this result in the equation giving the quantity $C_l bl$, we obtain:

$$C_l bl = \frac{16\pi}{9} \frac{r}{\lambda \sqrt{\lambda^2 + \frac{4}{9} \left(1 + \frac{2}{3\lambda} \tan \varepsilon\right)}}$$

In normal running, the value of $\tan \varepsilon = \dfrac{dR_d}{dR_l} = \dfrac{C_d}{C_l}$ is generally very low. For incidence angles near the optimal value and for ordinary aerofoils, $\tan \varepsilon$ is about 0.02. It follows that the preceding relation can be written as:

$$C_l bl = \frac{16\pi}{9} \frac{r}{\lambda \sqrt{\lambda^2 + \dfrac{4}{9}}}$$

The speed ratio at the blade tip and at the distance r from the rotor axis are respectively $\lambda_0 = \dfrac{\omega R}{V_1}$ and $\lambda = \dfrac{\omega r}{V_1}$. By eliminating ω and V_1 between this relation, we obtain $\lambda = \lambda_0 \dfrac{r}{R}$.

By carrying back the value of λ in the equation giving the quantity $C_l bl$, we get:

$$C_l bl = \frac{16\pi}{9} \frac{R}{\lambda_0 \sqrt{\lambda_0^2 \dfrac{r^2}{R^2} + \dfrac{4}{9}}}$$

3. APPLICATION TO THE BLADE CONFIGURATION. PRINCIPLES OF THE METHODOLOGY AND REMARKS

The tip-speed ratio λ_0 and the diameter of the windmill being known, the inclination angle I can be calculated for each value r from the relation:

$$\cot I = \frac{3}{2}\lambda = \frac{3}{2}\lambda_0 \frac{r}{R}$$

If the setting angle α is defined, the incidence angle is also determined $(i = I - \alpha)$. Then, from the aerofoil aerodynamic characteristics, the value of C_l may be obtained.

Thus for a given number of blades, the expression of $C_l bl$ enables us to determine without any difficulty, the chord of each blade as a function of the distance r to the rotation axis.

The expression giving $C_l bl$ shows that the chord of the blade at a given distance r from the rotor axis, decreases when the tip-speed ratio λ_0 increases. The rotors will then be lighter as they rotate more quickly.

For a given tip-speed ratio λ_0, the relationship shows also that the blade chord increases from the blade tip towards the hub. However this rule can be subjected to distortions. In some wind power plants, C_l does not keep a constant value along the blade, so the chord does not necessarily diminish from the tip towards the hub.

4. THEORTICAL AERODYNAMIC EFFICIENCY OF THE BLADE ELEMENT : OPTIMAL ANGLE OF INCIDENCE

The aerodynamic efficiency of the blade element situated within r and r + dr can be defined by the ratio :

$$\eta = \frac{dP_u}{dP_t} = \frac{\omega dM_1}{VdF_v} = \frac{UdF_u}{VdF_v}$$

dF_u and dF_v being respectively the projections of the aerodynamic force dR on the plane of rotation and on the rotor axis.

dP_u being the contribution of the blade element dr in the power provided by the rotor and dP_t the power supplied by the wind to the element dr.

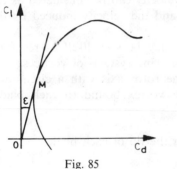

Fig. 85

As $dF_u = dR_l \sin I - dR_d \cos I$
$dF_v = dR_l \cos I + dR_d \sin I$

and $\cot I = \dfrac{U}{V}$, it follows that :

$$\eta = \frac{dR_l \sin I - dR_d \cos I}{dR_l \cos I + dR_d \sin I} \cot I$$

By putting $\tan \varepsilon = \dfrac{dR_d}{dR_l} = \dfrac{C_d}{C_l}$, we can write the previous expression under the form :

$$\eta = \frac{1 - \tan \varepsilon . \cot I}{\cot I + \tan \varepsilon} \cot I = \frac{1 - \tan \varepsilon . \cot I}{1 + \tan \varepsilon . \tan I}$$

The efficiency is as much higher as $\tan \varepsilon$ is lower. At the limit, if $\tan \varepsilon$ was equal to zero, the aerodynamic efficiency would be equal to the unity.

Actually the value of $\tan \varepsilon$ depends on the incidence angle. $\operatorname{Tan} \varepsilon$ is minimum for the angle of incidence corresponding to the point where the straight line OM becomes tangent to the Eiffel polar. For this particular value of incidence, the aerodynamic efficiency reaches a maximum.

II – *THE VORTEX THEORY OF GLAUERT*

The vortex theory has the merit of taking into account the induced rotation of the air flow which goes through the rotor.

1. THE VORTEX SYSTEM OF A WIND ROTOR

With blades whose length is limited, downstream of a wind rotor, there is for each blade a sheet of trailing vortices constituted for the major part by two vortices : one near the hub and the other at the blade tip.

Since each blade tip traces out a helix in the air flow as the turbine rotates, each trailing vortex will itself be of helical form.

It is the same thing with the vortex located near the hub which adds its action to the hub vortices of the other blades.

Moreover for determining the speed field, it is possible to substitute a bound vortex for the action of each blade. Thus, the vortex system of a wind rotor can be represented as seen in Figs. 86a and 86b.

At a given point in space, the wind velocity can be considered as the resultant of the undisturbed wind speed and the velocity induced by the vortex system.

The velocity induced by the vortex may be seen itself as resulting from the superimposition of the three following systems of vortices :

— the central vortex centered on the rotor axis with a circulation intensity $b\Gamma = \Gamma_0$ (Γ : circulation of the vortex bound to each blade, b : number of blades),

— the vortices bound to each blade,

— and the helical vortices shed from the tip of each blade.

2. THE CORRESPONDING ELECTRICAL CIRCUIT AND THE DETERMINATION OF THE INDUCED VELOCITIES

Fluid Mechanics teaches us that the velocity induced by vortex can be obtained by Biot's and Savart's laws or by Ampère's theorem in the same way as the magnetic fields created by direct current having the same shape as the considered vortices.

Thus, let us substitute in the preceding vortices system, the electrical circuit shown in fig. 86c consisting of :

— a central wire carrying a direct current $bI = I_0$,

— a set of b wires of length R arranged in the same manner as the blades of the windmill, each carrying a current I providing, in total, at their common point the current $bI = I_0$ to the former central wire,

— a set of helical wires, each supplying a current I to the previous wires.

For determining the magnetic field, the last set can be replaced by two other sets :

— The first one constituted by circular wires concentric to the rotor axis ;

— The second one, by meridian wires : each of them corresponding to a blade and leading a current I.

Let us calculate the magnetic fields created by these sets of wires on the assumption that the number of blades is infinite. In that hypothesis, for the set of b wires corresponding to the blades, can be substituted a conducting disc receiving, in its centre, the current I_0 and delivering the same current to the helical set at its perimeter.

Determine, at first, the magnetic field created by the central wire, the disc and the meridian wires in two planes perpendicular to the rotation axis and as near as possible to the disc, one being situated upstream and the second downstream of the disc representing the rotor.

The streamlines of the magnetic fields consist of circles concentric to the rotor axis and because of symmetry they occur upstream as well as downstream of the disc.

Downstream of the disc, the fields due to the various wires are additive. In front of the disc, it is not the same case.

It must be pointed out that the magnetic fields H_C and H_M created by the central wire and the meridian wires (theoretically an infinite number) keep a constant value at an equal distance from the axis, whatever position is considered :

Wind rotors vortex systems

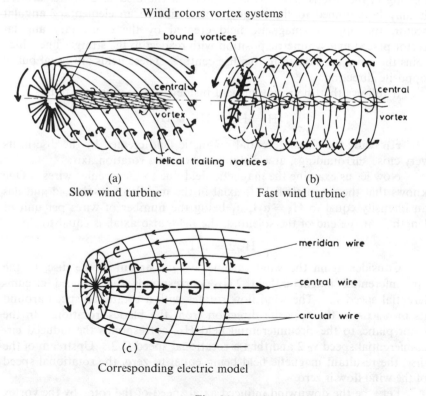

(a)
Slow wind turbine

(b)
Fast wind turbine

(c)
Corresponding electric model

Fig. 86

either in a plane in front of or behind the disc or in the plane of the disc itself, because of the close proximity of all these planes.

If we apply Ampere's theorem upstream of the disc along a circumference concentric to the rotation axis, we obtain zero for the intensity of the magnetic field because no current passes through the circular area. This means that the magnetic field H_D due to currents flowing in the disc is of an equal intensity but of an opposite direction to the magnetic fields created by the central wire and the meridian wires.

$$H_D = H_C + H_M$$

Downstream, these fields are additive. The resultant field at an equal distance r from the rotor axis, is then:

$$H = H_D + H_C + H_M = 2\,H_D$$

The magnetic fields created by the disc, the central and the meridian wires keep the same intensity owing to the close proximity of these planes.

Inside the surface of the disc itself, the field due to the electric current flowing in the disc is zero: At each point of the disc, as a matter of fact, it may be adjoined to the magnetic field due to an elementary angular sector, the opposite magnetic field created by the elementary angular sector placed in a symmetric position with respect to the straight line which joins the point considered to the disc centre. These fields are equal but of opposite direction.

It follows that the resultant field inside the disc area at a radius r, is:

$$H = H_C + H_M = H_D$$

H_D being the circumferential magnetic field created by the disc in its very close surroundings, at a distance r from the rotation axis.

Now let us examine the magnetic field due to the circular wires. One knows that the magnetic field is axial in the middle of a solenoïd and has an intensity equal to $H_S = n_1 I$, n_1 being the number of wires per unit of length. At the end of the solenoid, the field, also axial, is equal to:

$$H_S/2 = n_1 I/2.$$

Consider again the wind turbine. Downstream of the disc, to the circumferential magnetic field $2\,H_D$, corresponds the induced circumferential speed v_θ. The wind flow rotates with a rotational speed around its own axis in the opposite direction from the blades' rotation. In the rotor plane, to the circumferential field H_D, corresponds the induced circumferential speed $v_\theta/2$ and then a rotational speed $\Omega/2$. Upstream of the disc, the resultant magnetic field being equal to zero, the rotational speed of the wind flow is zero.

Let v be the downwind induced axial speed of the rotor by the vortex system. This speed, which corresponds to the above axial field $n_1 I$, is oriented in a direction opposite to the wind velocity V_1.

In the plane of rotation the induced axial speed rises only to $\frac{v}{2}$, the corresponding magnetic field being $\frac{n_1 I}{2}$.

In conclusion, the resultant axial wind velocity is:

$$V = V_1 - \frac{v}{2} \text{ through the rotor}$$

$$V_2 = V_1 - v \text{ downstream of the rotor.}$$

By eliminating v between these two equations, it is found that:

$$V = \frac{V_1 + V_2}{2}$$

So, the value of V obtained by Betz' theory, holds.

Downstream of the rotor, the rotational speed of the wind flow relative to the blade rises to $\omega + \Omega$.

Let $\omega + \Omega = h\omega$.

There follows: $\Omega = (h - 1)\,\omega$.

Under these conditions, the angular velocity of the air flow in the plane of the rotor with respect to the blades can be expressed as:

$$\omega + \frac{\Omega}{2} = \left(\frac{1 + h}{2}\right)\omega$$

At a distance r from the rotor axis, this corresponds to a circumferential speed:

$$U' = \left(\frac{1 + h}{2}\right)\omega r.$$

Let $V_2 = k V_1$; the axial speed through the rotor can be written as:

$$V = \frac{V_1 + V_2}{2} = \frac{1 + k}{2} V_1$$

The inclination angle and the relative speed W at radius r are then given in the rotor plane by the following relations:

$$\cot I = \frac{U'}{V} = \frac{\omega r}{V_1} \frac{1 + h}{1 + k} = \lambda \frac{1 + h}{1 + k} = \lambda_e$$

$$W = \frac{V_1(1 + k)}{2 \sin I} = \frac{\omega r(1 + h)}{2 \cos I}$$

3. AXIAL THRUST AND TORQUE CALCULATIONS

Consider the blade elements situated inside the interval r and r + dr. Calculate in two ways the forces to which these elements are subject:
— Firstly, by evaluating the aerodynamic action on the profiles,
— and then, by applying the fundamental laws of dynamics to the airflow between r and r + dr.

a) FIRST EVALUATION

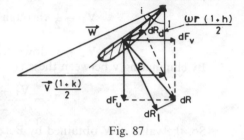

As previously, we can write:

$$dR_l = \frac{1}{2}\rho C_l W^2 l\, dr$$

$$dR_d = \frac{1}{2}\rho C_d W^2 l\, dr$$

Fig. 87

Projecting $\vec{dR}$ first upon the rotor axis and then upon the circumferential speed $\vec{U}$, we obtain:
— for the axial component of dR:

$$dFv = dR_l \cos I + dR_d \sin I = \frac{1}{2}\rho l W^2 dr\, (C_1 \cos I + C_d \sin I)$$

— and for the tangential component:

$$dF_u = dR_l \sin I - dR_d \cos I = \frac{1}{2}\rho l W^2 dr\, (C_l \sin I - C_d \cos I)$$

Taking into account the relation: $\tan \varepsilon = C_d/C_l$, these equations can be written as:

$$dF_v = \frac{1}{2}\rho l W^2 C_l \frac{\cos (I - \varepsilon)}{\cos \varepsilon} dr \text{ and } dF_u = \frac{1}{2}\rho l W^2 C_l \frac{\sin (I - \varepsilon)}{\cos \varepsilon} dr$$

The contribution of the blade elements situated inside the interval (r, r + dr) in the axial thrust is:

$$dF = b\, dF_v = \frac{1}{2}\rho b l W^2 C_l \frac{\cos (I - \varepsilon)}{\cos \varepsilon} dr$$

and in the areodynamic torque :

$$dM = rb\, dF_u = \frac{1}{2}\rho b l r W^2 C_l \frac{\sin (I - \varepsilon)}{\cos \varepsilon} dr.$$

b) SECOND EVALUATION

Determine now these two quantities by applying the general theorems of dynamics to the wind flow which goes through the rotor between r and r + dr.

Consider the axial momentum of the flow through the annulus. The thrust dF is equal to the product of the rate of mass flow m through the element with the change in the axial velocity i.e :

$$dF = m \, \Delta V = m \, (V_1 - V_2).$$

As : $$m = \rho 2\pi r dr \, V = \rho \pi r dr \, (1 + k) \, V_1$$

it follows that :

$$dF = \rho \pi r dr \, V_1^2 \, (1 - k^2)$$

In the same way, by considering the angular momentum, we obtain for the elementary torque dM :

$$dM = m \, \Delta \omega r^2 = m r^2 \Omega$$

where $\Delta \omega = \Omega$ is the change in angular velocity of the air on passing through the airscrew. Then :

$$dM = \rho \pi r^3 dr \, V_1 \, (1 + k) \, \Omega$$
$$dM = \rho \pi r^3 dr \, \omega V_1 (1 + k) \, (h - 1)$$

c) CONSEQUENCE

Comparing the value of dF determined above with that obtained by direct aerodynamic considerations leads, after substituting for W, its value as a function of V_1, to :

$$C_l bl = \frac{2\pi r V_1^2 \, (1 - k^2) \cos \varepsilon}{W^2 \cos (I - \varepsilon)} = \frac{8\pi r \, (1 - k) \cos \varepsilon \sin^2 I}{(1 + k) \cos (I - \varepsilon)}$$

In a similar manner, from equalling the expressions of dM, it is found that :

$$C_l bl = \frac{2\pi \omega r V_1 \, (1 + k) \, (h - 1) \cos \varepsilon}{W^2 \sin (1 - \varepsilon)} = \frac{4\pi r \, (h - 1) \sin 2I \cos \varepsilon}{(h + 1) \sin (I - \varepsilon)}$$

From these equations, it is possible, after some manipulation, to get the following ones :

$$G = \frac{1 - k}{1 + k} = \frac{C_l bl \cos (I - \varepsilon)}{8\pi r \cos \varepsilon \sin^2 I}$$

$$E = \frac{h - 1}{h + 1} = \frac{C_l bl \sin (I - \varepsilon)}{4\pi r \sin 2 I . \cos \varepsilon}$$

By dividing each member of the first equation above, by the second, we obtain:

$$\frac{G}{E} = \frac{(1-k)(h+1)}{(h-1)(1+k)} = \cot(I-\varepsilon)\cot I$$

4. LOCAL POWER COEFFICIENT

The maximum power capable of being extracted from the wind flow passing inside the annulus (r, r + dr) is given by the equation:

$$dP_u = \omega dM = \rho\pi r^3 dr\,\omega^2(1+k)(h-1).$$

This value corresponds to a local power coefficient:

$$C_p = \frac{dP}{\rho\pi r dr\,V_1^2} = \frac{\omega^2 r^2}{V_1^2}(1+k)(h-1) = \lambda^2(1+k)(h-1).$$

λ being equal to $\omega r/V_1$.

Maximum value of the local power coefficient for an ideal windmill

Determine the maximum value that can be reached by the local power coefficient.

For this purpose, consider an ideal windmill having an infinite number of blades without drag. As $C_d = 0$ for each profile, $\tan\varepsilon = C_d/C_l = 0$.
Under these circumstances, the equation giving G/E can be written as:

$$\frac{G}{E} = \frac{(1-k)(h+1)}{(h-1)(1+k)} = \cot^2 I = \frac{\lambda^2(1+h)^2}{(1+k)^2}$$

After simplification, there follows:

$$\lambda^2 = \frac{1-k^2}{h^2-1}$$

from which, it is found that:

$$h = \sqrt{1 + \frac{1-k^2}{\lambda^2}}$$

Setting this value of h in the equation of the power coefficient C_p leads to:

$$C_p = \lambda^2(1+k)\left(\sqrt{1 + \frac{1-k^2}{\lambda^2}} - 1\right)$$

For a given value of λ, the power coefficient has a maximum when:

$$\frac{dC_p}{dk} = 0$$

The calculation shows that the maximum is obtained for a value of k which satisfies the equation:

$$\lambda^2 = \frac{1 - 3k + 4k^3}{3k - 1}$$

This equation can be written as:

$$4k^3 - 3k\,(\lambda^2 + 1) + \lambda^2 + 1 = 0$$

Let

$$k = \sqrt{\lambda^2 + 1}\,\cos\theta$$

Substituting for k its value, in the previous equation gives, after dividing by $(\lambda^2 + 1)^{3/2}$:

$$4\cos^3\theta - 3\cos\theta + \frac{1}{\sqrt{\lambda^2 + 1}} = 0$$

As:

$$4\cos^3\theta - 3\cos\theta = \cos 3\theta \qquad \text{we can write:}$$

$$\cos 3\theta = -\frac{1}{\sqrt{\lambda^2 + 1}} \quad \text{i.e.} \quad \cos(3\theta - \pi) = \frac{1}{\sqrt{\lambda^2 + 1}}$$

from which there follows:

$$\theta = \frac{1}{3}\cos^{-1}\left(\frac{1}{\sqrt{\lambda^2 + 1}}\right) + \frac{\pi}{3} = \frac{1}{3}\tan^{-1}\lambda + \frac{\pi}{3}$$

For each value of λ, it is possible to determine θ then k, and therefore the maximum value of C_p.

5. OPTIMAL VALUES OF THE INCLINATION ANGLE AND THE QUANTITY C_1bl

We have obtained for the inclination angle I and the quantity $C_l$bl the equations:

$$\cot I = \lambda_e = \lambda\,\frac{1 + h}{1 + k}$$

$$C_l b1 = \frac{8\pi r\,(1 - k)\cos\varepsilon\,\sin^2 I}{(1 + k)\cos(I - \varepsilon)}$$

According to the results obtained in the former paragraph, the knowledge of the angle θ leads to the determination of k then h and consequently to those of λ_e and I.

To calculate the values of the quantity C_lb1, we shall again consider an ideal windmill without blade drag ($\varepsilon = 0$).

Under these circumstances, the expression $C_l bl/r$ may be written as:

$$\frac{C_l bl}{r} = \frac{8\pi(1-k)}{(1+k)}\frac{1}{\lambda_e\sqrt{\lambda_e^2+1}}$$

The above relationships enable us to determine the inclination angle and the quantity $C_l bl/r$ in order to be in the optimized operating conditions. The knowledge of these quantities is indispensable for fixing the blade chord and the setting angle at any radius r.

To facilitate the application of the above results to wind rotor designs, the quantities λ_e, k, h, C_p, $C_l bl/r$ and I have been computed (OPTI program) for different values of λ between 0.1 and 10. The values obtained are shown in table 10.

We have also established a diagram showing the variation of the quantities $C_l bl/r$ and I as a function of λ. This diagram enables us to determine quickly the geometrical characteristics to be given to the blades in order to permit the performances of the wind turbine to reach a maximum for a given tip-speed ratio.

Using the diagram does not present any difficulty. Once the speed ratios λ_0 for which the performances must be optimized, has been chosen, it suffices to follow the arrows drawn on the graph from the point corresponding to the ratio r/R characterizing the position of the considered section in the lower axes system, to obtain the quantities I and $C_l bl/r$ at radius r. The value of I and $C_l bl$ are respectively read on the left and right vertical axes. The oblique straight lines in the lower coordinate system are represented analytically by the expression: $\lambda = \lambda_0 r/R$.

Each straight line corresponds to a value of λ_0.

For $\lambda_0 = 7$, the monogram gives $I = 9°$ and $C_l bl/r = 0.3$ at a radius r = 0.6 R.

In the same way, taking into account that the speed ratio is:

$$\lambda = \lambda_0 r/R = 7 \times 0.6 = 4.2$$

in the considered section, we obtain from table 10 for $\lambda_0 = 4.2$, $I = 8°928$ and $C_l bl/r = 0.305$ values which are in close agreement with the results given by the graph.

If the incidence angle i is known, the lift coefficient C_l and the setting angle $\alpha = I - i$ can be determined. The number of blades b being given, the chord of the blade at radius r can be calculated.

So, the problem is reduced to the choice of the incidence angle.

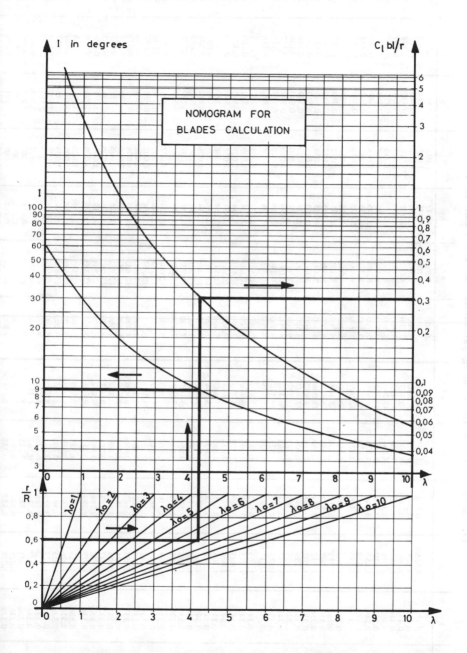

TABLE 10

Optimum values of the running parameters as a function of λ

λ	λe	k	h	c_p	$c_1 bl/r$	I
0.100	0.670	0.473	8.866	0.116	11.149	56.193
0.200	0.768	0.451	4.574	0.207	9.819	52.460
0.300	0.873	0.432	3.168	0.279	8.600	48.867
0.400	0.984	0.416	2.483	0.336	7.506	45.466
0.500	1.099	0.403	2.086	0.381	6.541	42.290
0.600	1.219	0.393	1.830	0.416	5.700	39.358
0.700	1.343	0.384	1.655	0.444	4.975	36.672
0.800	1.470	0.377	1.530	0.467	4.353	34.227
0.900	1.600	0.371	1.437	0.485	3.821	32.009
1.000	1.732	0.366	1.366	0.500	3.367	30.000
1.100	1.866	0.362	1.311	0.512	2.980	28.183
1.200	2.002	0.359	1.267	0.522	2.648	26.537
1.300	2.140	0.356	1.232	0.531	2.363	25.046
1.400	2.279	0.353	1.203	0.538	2.118	23.692
1.500	2.419	0.351	1.179	0.544	1.906	22.460
1.600	2.560	0.349	1.159	0.549	1.723	21.337
1.700	2.702	0.348	1.142	0.553	1.563	20.310
1.800	2.844	0.346	1.128	0.557	1.423	19.370
1.900	2.988	0.345	1.115	0.560	1.300	18.506
2.000	3.132	0.344	1.105	0.563	1.191	17.710
2.100	3.276	0.343	1.095	0.565	1.095	16.976
2.200	3.421	0.343	1.087	0.568	1.010	16.296
2.300	3.566	0.342	1.080	0.570	0.934	15.666
2.400	3.711	0.341	1.074	0.571	0.865	15.080
2.500	3.857	0.341	1.068	0.573	0.804	14.534
2.600	4.003	0.340	1.063	0.574	0.749	14.025
2.700	4.150	0.340	1.059	0.576	0.699	13.549
2.800	4.296	0.339	1.055	0.577	0.654	13.103
2.900	4.443	0.339	1.051	0.578	0.613	12.684
3.000	4.590	0.339	1.048	0.579	0.586	12.290
3.100	4.737	0.338	1.045	0.580	0.542	11.919
3.200	4.884	0.338	1.042	0.580	0.511	11.569
3.300	5.032	0.338	1.040	0.581	0.482	11.239
3.400	5.180	0.337	1.038	0.582	0.456	10.926
3.500	5.328	0.337	1.036	0.582	0.431	10.630
3.600	5.476	0.337	1.034	0.583	0.409	10.349
3.700	5.624	0.337	1.032	0.583	0.388	10.083
3.800	5.772	0.337	1.030	0.584	0.369	9.829
3.900	5.920	0.337	1.029	0.584	0.351	9.588
4.000	6.068	0.336	1.027	0.585	0.334	9.358
4.100	6.217	0.336	1.026	0.585	0.319	9.138
4.200	6.365	0.336	1.025	0.585	0.305	8.928
4.300	6.514	0.336	1.024	0.586	0.291	8.728
4.400	6.662	0.336	1.023	0.586	0.278	8.536
4.500	6.811	0.336	1.022	0.586	0.267	8.353
4.600	6.960	0.336	1.021	0.586	0.255	8.177
4.700	7.108	0.336	1.020	0.587	0.245	8.008
4.800	7.257	0.335	1.019	0.587	0.235	7.846
5.000	7.555	0.335	1.018	0.587	0.217	7.540
5.100	7.704	0.335	1.017	0.588	0.209	7.396
5.200	7.853	0.335	1.016	0.588	0.201	7.257
5.300	8.002	0.335	1.016	0.588	0.194	7.123
5.400	8.151	0.335	1.015	0.588	0.187	6.994
5.500	8.300	0.335	1.015	0.588	0.180	6.870
5.600	8.449	0.335	1.014	0.588	0.174	6.750
5.700	8.598	0.335	1.014	0.589	0.168	6.634
5.800	8.747	0.335	1.013	0.589	0.163	6.522
5.900	8.897	0.335	1.013	0.589	0.157	6.413
6.000	9.046	0.335	1.012	0.589	0.152	6.308
6.100	9.195	0.335	1.012	0.589	0.147	6.207
6.200	9.344	0.335	1.011	0.589	0.143	6.108
6.300	9.494	0.335	1.011	0.589	0.138	6.013
6.400	9.643	0.355	1.011	0.589	0.134	5.920
6.500	9.792	0.334	1.010	0.589	0.130	5.831
6.600	9.942	0.334	1.010	0.590	0.126	5.744
6.700	10.091	0.334	1.010	0.590	0.122	5.659
6.800	10.241	0.334	1.010	0.590	0.119	5.577
6.900	10.390	0.334	1.009	0.590	0.116	5.498
7.000	10.539	0.334	1.009	0.590	0.112	5.420
7.100	10.689	0.334	1.009	0.590	0.109	5.345
7.200	10.838	0.334	1.009	0.590	0.106	5.271
7.300	10.988	0.334	1.008	0.590	0.103	5.200
7.400	11.137	0.334	1.008	0.590	0.101	5.131
7.500	11.287	0.334	1.008	0.590	0.098	5.063
7.600	11.436	0.334	1.008	0.590	0.096	4.997
7.700	11.586	0.334	1.007	0.590	0.093	4.933
7.800	11.735	0.334	1.007	0.590	0.091	4.871
7.900	11.885	0.334	1.007	0.590	0.088	4.810
8.000	12.034	0.334	1.007	0.591	0.086	4.750
8.100	12.184	0.334	1.007	0.591	0.084	4.692
8.200	12.334	0.334	1.007	0.591	0.082	4.635
8.300	12.483	0.334	1.006	0.591	0.080	4.580
8.400	12.633	0.334	1.006	0.591	0.078	4.526
8.500	12.782	0.334	1.006	0.591	0.077	4.473
8.600	12.932	0.334	1.006	0.591	0.075	4.422
8.700	13.082	0.334	1.006	0.591	0.073	4.371
8.800	13.231	0.334	1.006	0.591	0.071	4.322
8.900	13.381	0.334	1.006	0.591	0.070	4.274
9.000	13.531	0.334	1.005	0.591	0.068	4.227
9.100	13.680	0.334	1.005	0.591	0.067	4.181
9.200	13.830	0.334	1.005	0.591	0.065	4.136
9.300	13.980	0.334	1.005	0.591	0.064	4.092
9.400	14.129	0.334	1.005	0.591	0.063	4.048
9.500	14.279	0.334	1.005	0.591	0.061	4.006
9.600	14.429	0.334	1.005	0.591	0.060	3.965
9.700	14.578	0.334	1.005	0.591	0.059	3.924
9.800	14.728	0.334	1.005	0.591	0.058	3.884
9.900	14.878	0.334	1.005	0.591	0.057	3.845

6. LOCAL POWER COEFFICIENT CAPABLE OF BEING REACHED BY USE OF IMPERFECT BLADES HAVING A NON NEGLIGIBLE DRAG.   OPTIMAL INCIDENCE ANGLE

Consider the blade elements situated between r and r + dr.
As previously, the local power coefficient is defined by the relation:

$$C_p = \frac{\omega \, dM}{\rho \pi r dr \, V_1^3} = \frac{V dF}{\rho \pi r dr \, V_1^3} \frac{\omega \, dM}{V dF} = \frac{V dF}{\rho \pi r dr \, V_1^3} \cdot \frac{U dF_u}{V dF_v}$$

Substituting for dF, dF_u, dF_v, V their values and taking into account the relations:

$$\cot I = \lambda \frac{1+h}{1+k} \text{ and } \tan \varepsilon = \frac{C_d}{C_l}$$

we obtain:

$$C_p = \frac{(1+k)(1-k^2)}{(1+h)} \cdot \frac{1 - \tan \varepsilon \cdot \cot I}{1 + \tan \varepsilon \cdot \tan I}$$

When $\tan \varepsilon = 0$ the first fraction of the second member represents the power coefficient of an ideal windmill at radius r.   We have seen that this coefficient reaches its maximum value when the conditions of table 10 are fulfilled.

In the hypothesis where $\tan \varepsilon$ is different from zero (blades having drag, usual case), figure 87b shows, as a function of λ, the maximum power coefficient capable of being reached for various values of the ratio C_d/C_l.

The graph demonstrates that obtaining good performances for high blade tip-speed ratio λ_0 requires blades having a very low roughness component.

For a given blade tip-speed ratio λ_0, the power coefficient is as much higher as $\tan \varepsilon$ is lower.   It reaches its maximum at the point of the Eiffel polar which corresponds to the minimum value of $\tan \varepsilon$ i.e. to the minimum value of the ratio C_d/C_l.

7. INFLUENCE OF THE NUMBER OF BLADES

The previous theory assumes that the number of blades is infinite. Actually, of course, it is limited.

There will follow losses of energy due to a greater concentration of vortices.   These losses of energy have been studied by Rohrbach, Worobel, Goldstein and Prandtl.

According to Prandtl, the reduction of efficiency which results, is given for a wind machine having b blades, by the relation:

$$\eta_b = \left(1 - \frac{1.39}{b} \sin I_0\right)^2$$

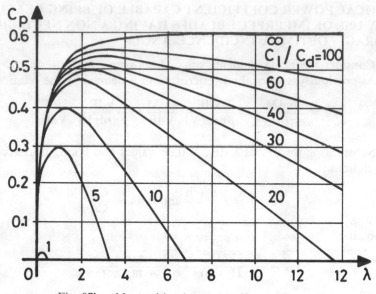

Fig. 87b – *Maximal local power coefficients* (b = x).

I_0 being the inclination angle at the blade tip.

In the hypothesis where the wind rotor is running in the neighbourhood of optimal conditions:

$$\sin I_0 = \frac{1}{\sqrt{1 + \cot^2 I_0}} = \frac{2}{3\sqrt{\lambda_0^2 + 4/9}}$$

Assuming that the Prandtl relation may be extended to these conditions, it follows that:

$$\eta_p = \left(1 - \frac{0.93}{b\sqrt{\lambda_0^2 + 0.445}}\right)^2$$

Let us point out that the original Prandtl relation was established strictly for low loaded airscrews.

In fact, the use of the above expression in the computational programs placed in the appendix leads, for wind rotors tested in a wind tunnel and normally loaded, to power coefficients in close agreement with the experimental results.

8. PRACTICAL DETERMINATION OF THE BLADE CHORD AND THE SETTING ANGLE

There is not just one method of choosing the angle of incidence and therefore of defining the geometry of the blades.

We can take of course, in each section, an angle of incidence equal to the optimal value i_0 which lowers the drag/lift ratio C_d/C_l in order to obtain higher efficiency. The angle of inclination being obtained from the graph or from table 7, the setting angle is determined by the relation :

$$\alpha = I - i_0$$

The lift coefficient C_l being also known, the chord of the profile at radius r can be readily deduced from the values of the quantity $C_l bl/r$.

In fact, this method is seldom completely applied. Effectively, if it is justified to take, near the blade tip, for the incidence angle, the optimal value because of the importance of the swept area per length unit of blade, the method leads, very often, to considerable values of chord in the neighbourhood of the rotor hub.

The lift coefficient corresponding to i_0 is generally not very high (about 0.9). By taking an incidence angle becoming higher towards the hub, holding, however the running representative point in the increasing part of the Eiffel polar which corresponds to value of tan ε lower than 0.1, we reduce the blade chord without decreasing significantly the efficiency because near the hub, the swept area per length unit of blade is small. For example, we shall choose a linear variation of the incidence angle which will vary from i_0 for r > 0.8 R to a higher value near the hub.

In practice, for usual aerofoils (Gottingen 623, NACA 4412, 4415, 4418, 23012, 23015, 23018) the maximum angle of incidence will be chosen to equal 10° or 12° in normal running at a distance 0.2 R from the rotor shaft. For such a value of the incidence, the lift coefficient is about 1.2, 1.3.

Another possibility consists in adopting a rotor of such shape that the quantity H maintains a constant value along the blade.

$$H = 2\pi r \tan \alpha = 2\pi R \tan \alpha_0$$

The setting angle α_0 at the blade tip is chosen equal to $I_0 - i_0$, I_0 being the inclination angle at the blade tip.

The knowledge of α which follows enables one to determine i, then C_1 and therefore, the chord l at any distance r from the rotor axis. It will be convenient, however, to verify that the incidence angle does not reach prohibited values at radius 0.2 or 0.3 R capable of compromising the wind turbine efficiency.

If the variation of the angle of incidence along the blade has been well chosen, the reduction of efficiency is slight because the swept area per length unit of blade decreases from the blade tip towards the hub (This reduction may be palliated easily by a slight increase of the rotor diameter in order to keep the power at a constant value.) On the other hand, the decrease of starting torque which follows the reduction both of the chord and the setting angle is more inconvenient. Nevertheless, this solution is often adopted because it leads to lighter wind rotor.

Another factor, which must be taken into consideration by the engineer, is that of manufacturing.

The blade shapes determined by applying the preceding methods are twisted. The difficulty of fabrication is reduced if fibreglass construction is used. The leading or the trailing edges can be made straight or not. But if steel or aluminium is used, it is very desirable that both the leading and the trailing edges should be straight.

In the appendix will be found several computational programs for blade calculation. One of them (EOLE program) relates to rotors having straight leading and trailing edges.

In the latter case, the wind rotor is firstly determined according to the classical method. Then two or three profiles are kept unaltered. The others are arranged so that their leading and trailing edges are in alignment with the leading and trailing edges of the profiles taken as reference. The calculation of wind turbine performances shows that rectifications of the leading and trailing edges lead to a reduction of the efficiency, but a reduction that is insignificant, if the reference profiles are judiciously chosen.

For testing the preceding vortex method, we have applied it to the blades of the aerogenerator of Nogent-Le Roi of 30.2 m diameter. The method has lead to the same values of chord calculated by the constructor L. Romani.

In chapter VIII will be found examples of designs of wind power plants calculated according to the above vortex method.

III − CALCULATIONS OF WIND TURBINE CHARACTERISTICS

The determination of the mechanical characteristics (torque and power coefficient as a function of λ_0) is particurlarly interesting. The knowledge of these characteristics before construction allows one, eventually, to make some modification in the initial design, in order to increase efficiency in the most economical manner.

But we must first recall the results previously reached.

In the preceding pages, we have shown that the axial thrust dF and the elementary torque dM of the aerodynamic forces acting on the blade elements situated between r and r + dr were equal to :

$$dF = \rho \pi r \ V_1^2 \ (1 - k^2) dr$$
$$dM = \rho \pi r^3 \omega V_1 \ (1 + k) \ (h - 1) dr.$$

Taking into account the expressions relating I and W to V and k to h :

$$W = \frac{V_1(1 + k)}{2 \sin I} = \frac{\omega r (1 + h)}{2 \cos I} \text{ and } \cot I = \lambda \frac{1 + h}{1 + k} = \lambda_0 \frac{r}{R} \frac{1 + h}{1 + k}$$

We then established the following relations:

$$G = \frac{1 - k}{1 + k} = \frac{C_l bl \cos (I - \varepsilon)}{8\pi r \cos \varepsilon . \sin^2 I} \quad \text{and}$$

$$E = \frac{h - 1}{h + 1} = \frac{C_l bl \sin (I - \varepsilon)}{4\pi r \sin 2I . \cos \varepsilon}$$

On integrating the preceding expressions of dF and dM, we obtain for the axial thrust and the moment:

$$F = \int_0^R \rho\pi V_1^2 (1 - k^2) r dr$$

$$M = \int_0^R \rho\pi V_1 \omega (1 + k) (h - 1) r^3 dr$$

and then for the axial thrust and moment coefficients:

$$C_F = \frac{2F}{\rho SV_1^2} = 2 \int_0^1 (1 - k^2) \frac{r}{R} d\left(\frac{r}{R}\right)$$

$$C_m = \frac{2M}{\rho SV_1^2 R} = 2 \int_0^1 \lambda(1 + k) (h - 1) \frac{r^2}{R^2} d\left(\frac{r^2}{R}\right)$$

Taking into account the relations:

$$\lambda = \frac{1 + k}{1 + h} \cot I \quad \text{and} \quad E = \frac{h - 1}{h + 1}$$

the previous one can be written as:

$$C_m = 2 \int_0^1 (1 + k)^2 . E \cot I . \frac{r^2}{R^2} . d\left(\frac{r}{R}\right)$$

Let $\quad f_r = (1 - k^2) \frac{r}{R}$ and $m_r = (1 + k)^2 E . \frac{r^2}{R^2} . \cot I$

we can write:

$$C_F = 2 \int_0^1 f_r d\left(\frac{r}{R}\right)$$

$$C_m = 2 \int_0^1 m_r d\left(\frac{r}{R}\right)$$

In optimal running, the values of k and h are known for each profile, so the calculation of axial thrust and torque is easy.

But, if the wind rotor rotates at a rotational speed which corresponds to a tip-speed ratio different from λ_0, the angles of incidence are not the same. The values of k and h, the axial thrust, the torque and then the coefficients C_F and C_m vary.

In practice to determine C_F and C_m, we shall consider several values of the ratio r/R (0.2, 0.4, 0.6, 0.8, 1 for instance).

To the corresponding radius, we shall make the angle of incidence vary arbitrarily, for instance, from degree to degree and we shall calculate for each value of the ratio r/R:

— the inclination angle I by adding to the considered incidence, the setting angle α (I = i + α),

— the values of C_l, ε,

— the quantities G, E, k and h,

— the tip-speed ratio λ_0 by using the relation: $\lambda_0 = \dfrac{R}{r} \dfrac{1 + k}{1 + h} \cot I$

— the quantity f_r and m_r.

TABLE 11

$\dfrac{r}{R}$						$\dfrac{r_1}{R_1}$		$\dfrac{r_2}{R_2}$
α						α_1		α_2
l						l_1		l_2
i		i_1	i_2	i_3	i_4	i_5		
C_l								
$\tan \varepsilon$								
I = α + i								
$G = \dfrac{1 - k}{1 + k} = \dfrac{C_l bl \cos (I - \varepsilon)}{8\pi r \cos \varepsilon \sin^2 I}$								
$k = \dfrac{1 - G}{1 + G}$								
$E = \dfrac{h - 1}{h + 1} = \dfrac{C_l bl \sin (I - \varepsilon)}{4\pi r \sin 2I \cos \varepsilon}$								
$h = \dfrac{1 + E}{1 - E}$								
$\lambda_0 = \dfrac{R}{r} \dfrac{1 + k}{1 + h} \cot I$								
$f_r = (1 - k^2) \dfrac{r}{R}$								
$m_r = (1 + k^2) E \cot I \cdot \dfrac{r^2}{R^2}$								

The calculations having been done, it is possible to draw:
— the variation curves of f_r and m_r as a function of λ_0, each curve corresponding to a given value of r/R,
— and then the variations of f_r and m_r as a function of λ_0, taken from the two previous graphs.

Determining the variation of the thrust, torque and power coefficients
C_f, C_m and C_p as a function of λ_0.

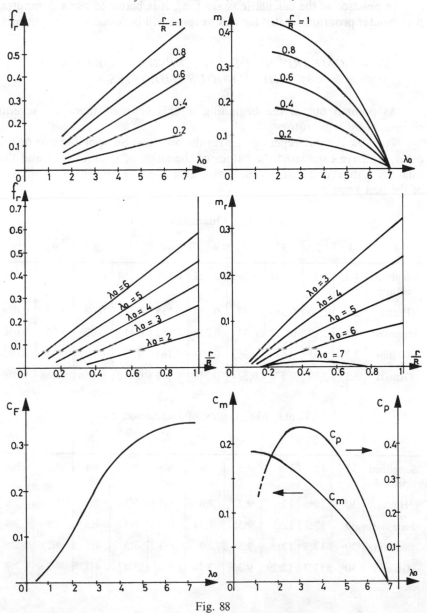

Fig. 88

The different values of the coefficients C_F and C_m as a function of λ_0 are obtained by measuring the areas situated between the various curves graduated in λ_0 values and the horizontal axes and then by multiplying the areas by two, according to the expressions of C_F and C_m.

In this chapter we have shown that the power coefficient C_p was related to C_m by the relation: $C_p = C_m \times \lambda_0$. Thus, determining the $C_p(\lambda_0)$ curve from the $C_m(\lambda_0)$ curve is easy.

In practice as the calculations are long, it is better to use a computer. A computer program created for this purpose will be found in the appendix.

IV – COMPARISON OF THE RESULTS OBTAINED BY THE VARIOUS THEORIES

As pointed out at the beginning of this chapter, there are several theories of the wind rotor.

To show the differences in the results obtained by applying one or the other, we have calculated the values of the angle of inclination I and the quantity $C_l bl/r$ under the most favourable running conditions, for each of the best known.

TABLE 12 : **Quantities $C_l bl/r$**

λ	1	2	3	4	5	6	7	8	9	10
Simplified method	4.65	1.32	0.608	0.345	0.222	0.155	0.114	0.087	0.069	0.056
Hütter	3.5	1.20	0.58	0.32	0.215	0.155	0.12	0.08	0.07	0.055
Stefaniak	3.37	1.19	0.572	0.33	0.216	0.152	0.114	0.088	0.069	0.056
Sabinin	3.71	1.31	0.63	0.363	0.237	0.167	0.125	0.097	0.076	0.062
Glauert	3.37	1.19	0.576	0.334	0.217	0.152	0.112	0.086	0.068	0.056

TABLE 13 : **Angles of inclination I**

λ	1	2	3	4	5	6	7	8	9	10
Simplified method	33°7	18°	12°	9°5	7°6	6°3	5°5	4°8	4°2	3°8
Hütter	30°	18°	12°	9°2	7°6	6°3	5°5	4°5	4°2	3°8
Stefaniak	30°	17°7	12°3	9°36	7°54	6°3	5°5	4°5	4°2	3°8
Sabinin	30°	17°5	12°	9°5	7°6	6°3	5°5	4°5	4°2	3°8
Glauert	30°	17°7	12°29	9°36	7°54	6°3	5°42	4°75	4°23	3°8

The values of the angle of inclination I and the quantity $C_l bl/r$ have been determined:

— for the simplified method, according to the relationships given at the beginning of this chapter,

— for the Stefaniak method from the table contained in the study "Windrad grösster Leistungabgabe" published in Forschung 19, Heft 1,

— for the Hütter theory, from the graph published in a paper presented by Hutter at the Meeting in Rome on New Sources of Energy,

— for the Sabinin theory from formulas taken from the book by I. Shefter "Wind-Powered Machines" translated from Russian into English and published by NASA.

The tables show that the divergence of the values of the quantity $C_l bl/r$ as of the angles of inclination, is small.

The values obtained by the Sabinin theory are however about 10 % greater than the values calculated by the other methods.

The simplified theory leads, for low values of λ, to values of $C_l bl/r$ and I slightly higher than those obtained by the other theories, but for $\lambda > 3$, the differences are quite insignificant.

B. Determination of the Blade Structure

In normal running, the blades must withstand the centrifugal forces, the bending moments caused by gusts and the gyroscopic effects which appear during the orientation changes which can accompany gusts. When the wind turbine is stopped facing the wind, the blades must also resist the most violent storms without breaking.

1. DETERMINATION OF THE BENDING STRESSES DUE TO GUSTS IN NORMAL RUNNING

The gusts produce an increment of the bending stresses in the blades consequent to the increase of wind velocity and to the direction variations.

As shown in ch. I, § 7, during gusts, wind velocity may change by 15 to 20 m/s within a second and wind direction may vary by several dozen degrees in the same time. So, it may occur that the blade is attacked under an unfavourable incidence. This effect has, as a consequence, an increase in the bending moment. The variation in the angle of incidence can cause a higher increment of stress than the relative increase of the wind speed.

It must be noted that the changes of wind direction which take place are essentially horizontal and very rapid.

On account of the machine inertia and the rapidity of the direction

change, the wind turbine axis cannot immediately reorientate to the wind
direction variations.   Thus, it may occur that the angle between the wind
turbine axis and the wind direction reaches up to 30° or 40° or even more.
Simultaneously, the wind velocity may be temporarily high.

Taking into account the shortness of the gust and its direction the
actual wind can very often reach the blades without slackening.

The more loaded blades are then those which are found in vertical
position and are moving against the actual wind.   Consider such a blade :
Let β be the angle between the wind turbine axis and the actual wind direction (see fig. 89).

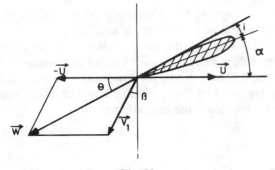

Fig. 89

The blade section at a distance r from the rotation axis, moving at
the circumferential speed $U = U_0 r/R$ is subjected to the relative wind
velocity :

$$W_1 = \sqrt{U^2 + V_1^2 + 2UV_1 \sin \beta}$$

The direction of this wind is inclined to the plane of notation by an
angle θ such as :

$$\tan \theta = \frac{V_1 \cos \beta}{U + V_1 \sin \beta}$$

The component of the aerodynamic force by length unit which acts
perpendicularly to the chord of the section is :

$$f_n = \frac{1}{2} \rho C_n l W^2$$

1 being the chord and C_n the aerodynamic coefficient of Lilienthal which
corresponds to an incidence angle : $i = \theta - \alpha$.

For given velocities U_0 and V_1, it is possible to calculate, by considering
several values of β, the maximum value which can be reached by the quantity $C_n W^2$ and consequently the maximum value of f_n in any section situated
at a distance r from the rotor axis.

The blades of the big aerogenerators usually have a small twist. This can be considered as negligible for the determination of the bending stresses. The values so obtained are slightly higher than the actual stresses but the difference is slight.

In the above hypothesis, the maximum bending moment due to the aerodynamic forces which act between the blade tip and the section situated at a distance x from the rotation axis rises to

$$M_x = \int_X^R (x - r) f_{nM} dr$$

f_{nM} being the maximum value of f_n.

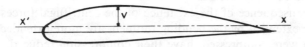

Fig. 90 – x x' Longitudinal, neutral axis.

Thus the bending stress in the section at the distance x from the blade root reaches the value:

$$\mathcal{R}_f = \frac{M_x v}{I_x}$$

where I_x represents the geometrical moment of inertia of the section relative to the longitudinal neutral axis of the section and v the distance from this axis to the more loaded fibre.

Due to the peculiar shape of the aerofoils usually used in the wind turbines and the usual disposal of the section, this fibre is generally an extrados fibre which is compressed.

The structure will be calculated in order to limit $\mathcal{R}_f$ at a given value which depends on the material used.

It must be specified that the obtained stresses are much higher than the normal running stresses which appear in the rated wind flow assumed to be uniform and parallel to the rotor axis.

If the blades have been aerodynamically determined according to the preceding methods, effective calculation shows that the maximum stresses take place not at the blade root but at a distance of approximately 0.6 R from the rotation axis.

2. STRESSES DUE TO CENTRIFUGAL FORCES IN NORMAL RUNNING

When the blades are perpendicular to the rotor shaft, the centrifugal force arising on the blade element, situated between the distance r and r + dr from the rotation axis, is given by the relation :

$$dF_c = \rho S \omega^2 r dr$$

S being the section of the material which constitutes the blade at the distance x.

Therefore the stress in this section due to the centrifugal force rises to :

$$\mathcal{R}_c = \frac{F_c}{S_x} = \frac{1}{S_x} \int_X^R \rho S \omega^2 r dr$$

This stress which corresponds to a tensile effort, comes in superposition to the bending stresses.

As a consequence of the existence of the centrifugal forces, the fibres which were extended by the bending moment are more extended and those ones which were compressed, have their compression reduced.

Let us specify that, in normal running, the stresses due to gravity forces are low compared to those created by centrifugal forces, these being themselves much less important than the bending stresses.

3. COMPENSATION OF BENDING BY CENTRIFUGAL FORCES IN THE NORMAL RUNNING

The stresses are reduced significantly in normal running when the blades are fastened on the hub so that the blades axes make with the rotation axis an angle lower than 90º, determined according to the relation which follows. The surface swept by the blades is then a conical one. This arrangement is generally called "raking" or "coning".

The moment of the aerodynamic force is counterbalanced by the

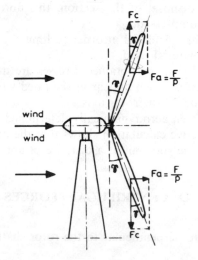

Fig. 91

moment of the centrifugal forces.   If the angle γ is well chosen, the resultant bending moment can be equal to zero, so that the blades are only subjected to tensile stresses.

Let F_a be the axial thrust, b the number of blades, F_c the centrifugal force which acts on each blade.   The ideal conditions are fulfilled when the angle γ seen in fig. 91 is such as :

$$\tan \gamma = \frac{F_a}{bF_c}$$

In normal running, in the absence of gust, the blade stresses are scarcely higher than those created by the centrifugal forces acting alone.

For a constant value of the angle γ, it must be noted that exact compensation cannot be observed except for a unique value of the rotational speed.

To solve the problem, in some wind power plants, the blades are hinged on the hub.   The angle γ varies in such a manner that it counterbalances the bending moment.   The " coning " is variable and the blades can " cone " independently of one another.

But whatever arrangement is adopted, it must be emphasized that, in normal running, the blades are not completely insulated from the bending due to gusts.   Moreover, the blades must be able to withstand the bending moment generated when the wind turbine is stopped facing the wind.

Consequent to the reduction of stress, it is possible according to Louis Vadot, to build raked wind machines having specific rotors up to 10 or 11.

With blades perpendicular to the axis, the limit is 9.

The raking improves the safety of the fast wind turbines especially when they have fixed blades.

4. GYROSCOPIC EFFECTS

The changes in the orientation of the rotor shaft create gyroscopic efforts.

Let ω be the rotational speed of the rotor shaft assumed to be constant and Ω its orientation speed around the vertical axis.

Consider a blade whose moment of inertia relative to the rotation axis is I_1 and two mobile trihedrons :

— The first one Oxyz(R) turning with the rotor shaft around the vertical axis, so that the vectors $\overline{\omega}$ and $\overline{\Omega}$ are brought by Ox and Oz axes, Oz being vertical.

— The second one Oxy'z' (R') carried away by the rotor, Oy', being directed along a blade axis.

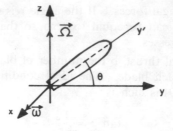

Fig. 92a

Assuming the moment of inertia of the blade relative to Oy' is negligible (and this is true within a reasonable approximation), the matrix of inertia and the moment of momentum of the blade with respect to the coordinate system Oxy'z' are given by the expressions which follow:

$$\square = \begin{bmatrix} I_1 & 0 & 0 \\ 0 & 0 & 0 \\ 0 & 0 & I_1 \end{bmatrix}$$

$$\square \, (\vec{\Omega} + \vec{\omega}) = \begin{bmatrix} I_1 & 0 & 0 \\ 0 & 0 & 0 \\ 0 & 0 & I_1 \end{bmatrix} \begin{bmatrix} \omega \\ \Omega \sin \theta \\ \Omega \cos \theta \end{bmatrix} = \begin{bmatrix} I_1 \, \omega \\ 0 \\ I_1 \, \Omega \cos \theta \end{bmatrix}$$

The principle of moment of momentum applied to the blade whose longitudinal axis coincides with Oy' with respect to Oxy'z', allows us to write:

$$\frac{d}{dt} \left[\square (\vec{\Omega} + \vec{\omega}) \right]_{R'} + (\vec{\Omega} + \vec{\omega}) \wedge \square \, (\vec{\Omega} + \vec{\omega}) = \vec{M}$$

$\vec{M}$ being the resultant moment applied to the blade with respect to the point 0, i.e. the resultant of the moment M_a of the aerodynamic forces and the embedding moment $\vec{M}_c$ due to the hub whose radius is assumed to have a negligible value.

All calculations done, we obtain:

$$\vec{M} = \begin{bmatrix} \dfrac{I_1}{2} \, \Omega^2 \sin 2\theta \\[2ex] 0 \\[2ex] I_1 \, \dfrac{d\Omega}{dt} \cos \theta - 2 I_1 \, \Omega\omega \sin \theta \end{bmatrix}$$

When the direction of the rotation axis remains fixed ($\Omega = 0$), the moment $\vec{M}$ is equal to zero.

We have then: $\vec{M} = \vec{M}_e + \vec{M}_a = 0$ i.e. $\vec{M}_e = - \vec{M}_a$

The embedding moment counterbalances the aerodynamic moment.

When the orientation of the rotation axis changes, the moment M which appears, creates supplementary stresses which increase the stresses in the material constituting the blades. The preceding equation which can be written as:

$\vec{M}_i + \vec{M}_e + \vec{M}_a = 0$ shows that $\vec{M}_e = - (\vec{M}_a + \vec{M}_i)$, the moment $\vec{M}_i$ being equal to $- \vec{M}$.

On the assumption that the orientation of the rotor axis is oscillating according to the following angular variation:

$$\alpha = \alpha_0 \sin 2\pi \frac{t}{T} = \alpha_0 \sin \omega_1 t, \text{ we can write:}$$

$$\Omega = \frac{d\alpha}{dt} = \alpha_0 \omega_1 \cos \omega_1 t = \Omega_0 \cos \omega_1 t$$

$$\frac{d\Omega}{dt} = - \Omega_0 \omega_1 \sin \omega_1 t = - \alpha_0 \omega_1^2 \sin \omega_1 t$$

ω_1 being smaller than ω, the most dangerous component for the blade of the moment $\vec{M}$ is the component along Oz' because it has the highest magnitude and acts perpendicularly to the rotation plane.

It includes two terms of unequal importance which are in quadrature:

$$I_1 \frac{d\Omega}{dt} \cos \theta = - I_1 \alpha_0 \omega_1^2 \sin \omega_1 t \cos \theta \text{ and}$$

$$- 2I_1 \Omega \omega \sin \theta = - 2I_1 \alpha_0 \omega_1 \omega \cos \omega_1 t \sin \theta$$

As ω_1 is small compared to ω, this last term which represents the moment of the Coriolis force, is the higher one. Its amplitude is at its maximum when $\theta = \pm \frac{\pi}{2}$.

The stress due to the gyroscopic effect is then at its maximum when the blade is vertical.

The quantity $2I_1 \alpha_0 \omega_1 \omega = 2I_1 \Omega_0 \omega$ represents the supplementary maximum bending moment due to the gyroscopic effect. It must be added to the bending moment due to the aerodynamic forces.

Let i_1 be the moment of inertia of the part of the blade situated between the blade tip and the distance x from the rotor axis, with respect to the longitudinal neutral axis of the section x. The maximum stress due to the gyroscopic effect at the distance x from the blade foot is given by the relation :

$$\mathcal{R}g = \frac{2 i_1 \Omega_0 \, \omega v}{Ix}$$

Resultant moment acting on the whole rotor

Let us calculate the resultant moment acting on the whole rotor when the orientation of the rotation axis is changing :

On resolving the moment $\vec{M}$ in its three components in the trihedron Oxyz, we obtain all calculations done :

$$\vec{M} = \begin{bmatrix} \dfrac{I_1}{2} \Omega^2 \sin 2\theta \\[2ex] 2I_1\omega\Omega \sin^2 \theta - \dfrac{I_1}{2} \dfrac{d\Omega}{dt} \sin 2\theta \\[2ex] - I\omega\Omega \sin 2\theta + I_1 \dfrac{d\Omega}{dt} \cos^2 \theta \end{bmatrix}$$

First let us consider a two-bladed wind machine. Let ωt be the angle of the first blade with Oy. The angle of the second blade with Oy is $(\omega t + \pi)$. The moment of inertia I of the whole rotor with respect to Ox is $I = 2I_1$. By superposing the moments we obtain :

$$\Sigma \vec{M} = \begin{bmatrix} \dfrac{I}{2} \Omega^2 \sin \omega t \\[2ex] 2I\omega\Omega \sin^2 \omega t - \dfrac{I}{2} \dfrac{d\Omega}{dt} \sin^2 \omega t \\[2ex] - I\Omega\omega \sin 2\omega t + \dfrac{d\Omega}{dt} \cos^2 \omega t \end{bmatrix}$$

Consider now a three-bladed machine. If ωt is the angle between the first blade and Oy, the angles of the other blades will be respectively :

$$\left(\omega t + \frac{2\pi}{3} \right) \quad \text{and} \quad \left(\omega t + \frac{4\pi}{3} \right)$$

By superposing the different moments and taking into account the equality $I = 3I_1$, we obtain for the three-bladed machines :

$$\Sigma \vec{M} = \begin{bmatrix} 0 \\[1ex] I \; \omega \; \Omega \\[1ex] I \; \dfrac{d\Omega}{dt} \end{bmatrix}$$

The three-bladed rotor appears then better balanced in its whole set than the two-bladed one. We should obtain a similar result for a wind turbine having a higher number of blades. It must be pointed out however that each blade is submitted separately to the gyroscopic effects in their entirety even if this blade is a part of a three-bladed machine.

Relative influence of the gyroscopic effects upon the blade stresses

The gyroscopic effects increase the stresses in the blades. The relative increment is all the more important as the tip-speed ratio is low.

M. Louis Vadot who designed the French aerogenerators erected in St Remy des Landes, points out that, at equal orientation speed, the gyroscopic stress varies in a 12 m/s wind, from 0.4 times the normal stress for $\lambda_0 = 5$ to 0.066 for $\lambda_0 = 10$, and 0.017 for $\lambda_0 = 15$.

5. VIBRATIONS

Among the causes of fatigue, vibrations are the main ones. They have various origins. They are caused especially by :

— The fact that the rotor shaft cannot be exactly oriented towards the wind direction. If the rotor axis is inclined with respect to this direction, the speed vectors triangles are not the same for each blade, due to the axial dissymmetry.

— The unequal distribution of the wind speeds on the area swept by the rotor (difference between the wind speeds at the higher and the lower part).

— The instantaneous variations of the wind speed.

— The gravity forces acting on each blade and whose direction is continuously varying with respect to the blade axis when the rotor is revolving.

— The variations in the wind direction which create changes of orientation of the rotor shaft and then make gyroscopic stresses appear inside the blades.

For all these reasons, and because blades are elastic structures, vibrations are always present. Their level must remain low enough not to compromise the safety of the wind plant.

The drawback of vibrations is that they create fatigue. By fatigue is meant the reduction of the material strength when subjected to alternative loads.

Figure 92b represents the Wöhler curve for a steel sample subjected to rotating bending. It gives the number of cycles which cause failure for a given maximal vibratory stress.

To limit the amplitude of the vibrations, effects of resonance between the rotor and the tower structures must be absolutely avoided.

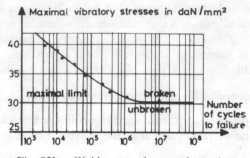

Fig. 92b – *Wohler curve for a steel sample*
(material fatigue characterization).

Blades

The blades are subjected to a forcing vibration with a frequency
of once per revolution, especially if the rotor revolves downwind of the
support. After their first passage behind the support, they vibrate at
their natural period until, once again, the tower shadow intervenes and
the cycle is repeated. To avoid damage, the natural frequency of blade
oscillations must be different from the number of revolutions per second
and not be an exact multiple of this number.

The amplitude of blade vibrations is greater when the rotor rotates
downwind of the support than when it is placed upwind.

The ratio of the amplitude downwind/upwind can reach two for a
support constituted by a lattice tower and four, for a concrete tower (Morrison : Testing of a Wind Power Plant).

Hence, placing the rotor upwind of the tower reduces fatigue inside
the blades.

Support

For a two-bladed rotor, the tower is stimulated twice per revolution
because of the difference in the mean wind speed in the upper and lower
part of the swept area. If the number of revolutions per second is n,
the frequency of the stimulating force on the tower (which is here the axial
thrust) is 2n. The support's first-mode natural frequency must be selected to be sufficiently displaced from the primary forcing frequency 2n so
as not to resonate. Care must also be taken to avoid higher-mode resonance.

If the natural frequency of the support is greater than 2n, it is referred
to as "stiff". Between 1n and 2n, the tower is characterized as "soft"
and below 1n as "very soft". The stiffer the tower, the heavier and more
costly it will be. Recent designs around the world use "soft" and "very
soft" supports.

The effects of vibrations in the tower are also reduced when the fixation of the blades on the hub is not rigid, especially when the blades are hinged on the hub (variable coning) or when the rotor can teeter by a small angle of up to 5 degrees in and out of the plane of rotation. Such arrangements reduce the fatigue caused by gusts and wind shear in the rotor shaft and the support, but they have the disadvantage of being more complicated.

6. DETERMINATION OF THE BENDING STRESSES DURING A STORM, WHEN THE ROTOR IS STOPPED FACING THE WIND

Let C_d be the drag coefficient of the blade when placed perpendicular to the wind. The force exerted on the blade element situated between r and r + dr is:

$$dF = \frac{1}{2} \rho C_d \, l V^2 \, dr = f_v \, dr$$

The bending moment at the distance x from the rotor axis is:

$$M_f = \int_X^R (r - x) \, f_v \, dr$$

In practice, the blade surface being almost perpendicular to the wind, the drag coefficient is approximately equal to 2.

When the setting angle α is relatively important, we can take for C_d the value:

$$C_d = 2 \cos \alpha.$$

The knowledge of the bending moment at distance x from the axis, allows us to determine the corresponding stress.

$$\mathcal{R}_f = \frac{M_f v}{I_x}$$

M_f being the above bending moment I_x and v having the same meaning as in the former first paragraph.

For information, note that the blades of the French aerogenerators of Nogent Le Roi and St Remy des Landes described in chapter VII, were calculated to withstand 63 m/s wind speeds at standstill with the rotor facing the wind.

In practice, the stresses at rest with the rotor facing the wind, prevail for the structure calculations of the wind turbines having a specific ratio λ_0 less than 4 and for the raked fast wind turbines, whatever their specific ratio may be.

For the high speed wind turbines having their blades perpendicular to the rotor shaft, the stresses, when stopped, in very strong winds, are lower than the stresses in normal running with gusts.

7. STRESSES DEVELOPED IN GEOMETRICALLY SIMILAR WIND ROTORS RUNNING IN A GIVEN WIND VELOCITY

Consider two geometrically similar rotors running in a given wind speed and revolving at the same tip-speed ratio λ_0. Assume them to be made of the same materials.

Give the index 1 to the rotor 1 parameters and the index 2 to the rotor 2 parameters.

Then, determine the stresses in corresponding sections of the rotors 1 and 2 due to the centrifugal forces and to the bending moment.

a) CENTRIFUGAL FORCE

The intensity of this force is given by the following expressions.

Rotor 1.

$$F_1 = \int_{x_1}^{R_1} \rho \omega_1^2 r_1 S_1 \, dr_1$$

ω_1 being connected to λ_0 by the relation:

$$\lambda_0 = \frac{\omega_1 R_1}{V} \qquad \text{i.e. } \omega_1 = \frac{\lambda_0 V}{R_1}$$

whence for F_1, the value:

$$F_1 = \int_{x_1}^{R_1} \rho \frac{\lambda_0}{R_1^2} V^2 r_1 S_1 \, dr_1 = \rho \frac{\lambda_0^2}{R_1^2} V^2 \int_{x_1}^{R_1} r_1 S_1 \, dr_1$$

Rotor 2

In the same way, we find:

$$F_2 = \rho \frac{\lambda_0^2}{R_2^2} V^2 \int_{x_2}^{R_2} r_2 S_2 \, dr_2$$

Thus we obtain for the stresses due to centrifugal forces in corresponding sections:

Rotor 1

$$\mathcal{R}_1 = \frac{F_1}{S_{x1}} = \rho \frac{\lambda_0^2}{R_1^2} \frac{V^2}{S_{x1}} \int_{x_1}^{R_1} r_1 S_1 \, dr_1$$

Rotor 2

$$\mathcal{R}_2 = \frac{F_2}{S_{x2}} = \rho \frac{\lambda_0^2}{R_2^2} \frac{V^2}{S_{x2}} \int_{x_2}^{R_2} r_2 S_2 \, dr_2$$

Taking into account the similarity of the two rotors, we can write:

$$\frac{1}{R_1^2 S_{x1}} \int_{x_1}^{R_1} r_1 S_1 dr_1 = \frac{1}{R_2^2 S_{x2}} \int_{x_2}^{R_2} r_2 S_2 dr_2$$

and therefore:
$$\mathscr{R}_1 = \mathscr{R}_2.$$

Hence in corresponding sections, the stresses due to centrifugal forces are equal.

b) BENDING MOMENT

In the former chapter, we have shown that the incidence angles at corresponding sections were equal for rotors running in similar conditions.

It results that, at equal wind speeds, the lift and the drag forces acting on corresponding elements are proportional to the square of measurement.

$$F_1' - k R_1^2 \qquad F_2' = k R_2^2$$

The bending moments at corresponding profiles are therefore proportional to the cube of measurements. It follows from this fact that:

$$\mathscr{R}_1' = \frac{M_1 v_1}{I_1} = \frac{k' R_1^3}{k'''} \frac{k'' R_1}{R_1^4} = \frac{k' k''}{k'''}$$

$$\mathscr{R}_2' = \frac{M_2 v_2}{I_2} = \frac{k' R_2^3}{k'''} \frac{k'' R_2}{R_2^4} = \frac{k' k''}{k'''} \qquad \text{thus } \mathscr{R}_1' = \mathscr{R}_2'$$

In corresponding profiles of geometrically similar wind turbines running in aerodynamic similitude in a given wind speed, the stresses due both to centrifugal forces and to the bending moment, are equal. It is likewise for the stresses resulting from the superposition of the two effects.

8. COMPARISON OF THE STRESSES PRODUCED IN TWO-BLADED, THREE-BLADED AND FOUR-BLADED GEOMETRICALLY SIMILAR WIND TURBINES HAVING THE SAME DIAMETER AND DESIGNED FOR A SAME TIP-SPEED RATIO

The shapes of the blades of the different turbines are assumed to be identical.

We have shown previously that the net section of the material constituting the blades is determined by the bending moment arising in normal running or at stoppage when the machine is facing a strong wind.

The relationships established at the beginning of this chapter have shown that the quantity $C_l bl$ depends only on the distance r to the rotation axis. The incidence angles and therefore the values of C_l being the same at an equal distance r from the rotor axis, the quantities bl are identical whatever the number of blades may be.   So, the width of each blade varies inversely to the number of blades.

Hence, at equal radius r, the chords of the blades of the two-bladed, three-bladed and four-bladed rotors are such as :

$$2l_2 = 3l_3 = 4l_4$$

The thicknesses of the blades are of course, in the same ratio.

The relative speeds being equal for corresponding sections, the bending moments arising on each blade are proportional to the chord then to the inverse of the number of blades.

$$2M_2 = 3M_3 = 4M_4.$$

Let us determine the stresses in the considered wind machines on the assumption that the blades are at first solid then hollow and made of the same materials.

a) SOLID SECTIONS

The maximum bending stresses at the distance r from the rotor axis rise to :

$$\mathcal{R}_2 = \frac{M_2 v_2}{I_2} \text{ for the two-bladed wind rotor}$$

$$\mathcal{R}_3 = \frac{M_3 v_2}{I_3} \text{ for the three-bladed rotor}$$

$$\mathcal{R}_4 = \frac{M_4 v_4}{I_4} \text{ for the four-bladed rotor.}$$

As $2v_2 = 3v_3 = 4v_4$ and $I_2 = \left(\dfrac{3}{2}\right)^4 I_3 = \left(\dfrac{2}{1}\right)^4 I_4$

all calculations effected :

$$\mathcal{R}_2 = 0.45 \ \mathcal{R}_3 = 0.25 \ R_4.$$

Moreover, geometrical calculations show that the mass of the two-bladed wind turbine is respectively equal to 1.5 times and 2 times the masses or the three and four-bladed wind turbines.

b) HOLLOW BLADES

We shall examine three cases :

1) The blades are hollow and the thicknesses of the material are in the same ratio as the width of the blades.

We find again the same result.

2) The blades are hollow and the thicknesses e_2, e_3, e_4 of the resistant structure are equal.

We find then :

$$\mathcal{R}_2 = \frac{2}{3} \mathcal{R}_3 = \frac{1}{2} \mathcal{R}_4$$

3) The blades are hollow and the stresses are equal.

$$\mathcal{R}_2 = \mathcal{R}_3 = \mathcal{R}_4$$

We obtain then : $\qquad e_2 = \frac{2}{3} e_3 = \frac{1}{2} e_4.$

A geometrical calculation shows that the weights P of the wind machines are proportional to the number of blades :

$$P_2 = \frac{2}{3} P_3 = \frac{1}{2} P_4$$

This proposition is general. At equal bending stress in the above conditions, the weight of the rotors are proportional to the number of blades.

Therefore, if we consider only the bending stress, it follows that a two-bladed rotor is more advantageous than a three-bladed one. However, some manufacturers find fault with it ; especially as it has a lower aerodynamic efficiency because of its greater tip losses and its vulnerability to vibration. It is not as well balanced as a three-bladed wind rotor for gyroscopic effects. When it is rotating, a two bladed propeller presents a cyclic variation of its moment of inertia about the pintle axis which passes successively through a maximum and through zero, twice per revolution. This may give rise to vibration.

Moreover, as a consequence of the number of blades and the difference between the wind velocity at the lower part and the higher part of the swept area, there is a cyclic variation of the reaction of the rotor on its foot-step bearings. For a downwind two-bladed wind rotor, this effect is much more accentuated as one of the blades crosses the tower shadow, while, at the same time, the other blade is subjected to the highest wind velocity at the upper part of the swept area. Thus the pitching moment is higher and the rotor shaft is subjected to greater bending stresses if the blades are rigidly fastened onto the hub.

In addition, the unequal distribution of wind speed between the lower and the upper part of the swept area creates a torque which has a tendancy to make the machine oscillate around its orientation axis. This torque has two maxima and two minima for a two-bladed propeller and three maxima and three minima per revolution for a three-bladed one. For a given wind gradient and for a given diameter of rotor and speed, according to Louis Vadot, the amplitude of the variation of this torque, about the pintle axis, is ten times greater with the two-bladed rotor than with the three-bladed one.

Nevertheless, two-bladed wind rotors are preferred by the manu-facturers in the USA, in Sweden and in Great Britain. On the other hand, in France and in Denmark, all the important wind installations use three-bladed wind rotors. However, in France, in low power plants, the two-bladed rotors are very common.

9. VARIATION OF STRESSES IN NORMAL RUNNING DUE TO CENTRIFUGAL FORCES AND TO THE BENDING MOMENT IN WIND MACHINES HAVING THE SAME DIAMETER AND THE SAME NUMBER OF BLADES BUT DESIGNED FOR DIFFERENT SPECIFIC RATIOS λ_0

At the beginning of this chapter, the single theory which is in good agreement with the vortex theory for tip-speed ratios higher than 3 has enabled us to establish the relationship:

$$C_l \mathrm{bl} = \frac{16\pi}{9} \frac{R}{\lambda_0 \sqrt{\lambda_0^2 \frac{r^2}{R^2} + \frac{4}{9}}}$$

Assume that in normal running in a wind velocity V, the incidence angles are equal at an equal distance from the rotor axis. If the blades are well designed, their widths at the distance r from the rotation axis are proportional to $1/\lambda_0^2$. This approximation is all the better as the term 4/9 is low compared to the term $\lambda_0^2 r^2/R^2$.

The thickness of the blades is proportional to the chord at equal radius r. Therefore the section of the blade is approximately proportional to $1/\lambda_0^4$.

Admitting this result, let us examine the variations of the stresses due to centrifugal forces and to the bending moments.

a) STRESS DUE TO CENTRIFUGAL FORCES

The expression:

$$\mathscr{R}_c = \frac{1}{S_x} \int_x^R \rho \omega^2 r S \, dr = \rho \frac{\lambda_0^2}{R^2} \frac{V_1^2}{S_x} \int_x^R r S \, dr$$

which gives the stress due to centrifugal forces in a section at the distance r from the rotor axis, shows that $\mathscr{R}_c$ is proportional to λ_0^2 for a given wind velocity V_1.

b) STRESS DUE TO THE BENDING MOMENT

This stress is given by the relation:

$$\mathscr{R}_f = \frac{Mv}{I_x}$$

If the incidence angles have the same values in the corresponding sections, the elementary aerodynamic forces are proportional to lW^2 i.e. to $lV_1^2(1 + \lambda^2) = lV_1^2(1 + \lambda_0^2 r^2/R^2)$.

If λ_0 r/R is high enough, compared to one, the term $\lambda_0^2 r^2/R^2$ is predominant. Hence, in the first approximation, the bending moment is proportionial to $l\lambda_0^2 V_1^2$. As l is proportional to $1/\lambda_0^2$, the value of M remains approximately constant.

The distance v from the neutral fibre to the more loaded fibre in the considered section, is proportional to $1/\lambda_0^2$.

The moment of inertia I_x is proportional to $1/\lambda_0^8$.

Therefore the stresses due to the bending moment must be proportional to λ_0^8. Indeed, the exponent is slightly lower.

Mr. Louis Vadot has determined the stresses in normal running for wind turbines operating in a 12 m/s wind.

In the absence of gusts, he obtained (see « La Houille Blanche », n° 5, 1958) the following values :

for $\lambda_0 = 5$ $\mathcal{R}$ max =   49 daN/cm^2
for $\lambda_0 = 10$ $\mathcal{R}$ max = 2 130 daN/cm^2
for $\lambda_0 = 15$ $\mathcal{R}$ max = 19 000 daN/cm^2

These maxima were observed at the distance r = 0.6 R.
These values correspond to a variation proportional to λ_0^n with $5 < n < 6$.

c) VARIATION OF THE STRESSES AT STOPPAGE AS A FUNCTION OF THE RATED TIP-SPEED RATIO

At standstill, the stresses are only due to the bending moment. The quantity lW^2 becomes lV^2, V being the velocity of the strongest wind the rotor must withstand.

The chord l is proportional to $1/\lambda_0^2$ and therefore, the bending moment also. As v and l vary proportionally to $1/\lambda_0^2$ and $1/\lambda_0^8$, the bending stress is proportional to λ_0^4.

For information, let us give some values of bending stresses at distance r = 0.6 R as determined by Louis Vadot, for a three-bladed wind rotor having its blades perpendicular to the rotating axis and operating in a 63 m/s wind.

for $\lambda_0 = 5$ R =   60 daN/cm^2
for $\lambda_0 = 10$ R =   800 daN/cm^2
for $\lambda_0 = 15$ R = 4 000 daN/cm^2

Indeed, the variation is proportional to λ_0^n, n being a figure within the interval 3 to 4.

For wind rotors having the blades perpendicular to their axis, the stresses in normal running prevail. But for wind turbines whose blades sweep a conical surface when rotating, the stresses are the highest when the rotor is at rest, facing a strong wind.

Moreover, it must be noted that the ratio of the stresses at standstill to the stresses in normal running varies, for wind rotors having blades perpendicular to their axes, as $1/\lambda_0^2$. This shows that the stresses of the blades under the action of the strongest wind have a more relative importance for slow wind turbines that for fast wind turbines characterized by high values of λ_0.

C. Determination of the Forces Acting on the Whole Plant

1. THE AXIAL THRUST

The knowledge of the chord and the setting angles enables us to evaluate the axial thrust.

What values can the axial thrust reach?

The measurements carried out on the Gedser windmill in Denmark have shown that the axial thrust by square meter of swept area, in normal running, was approximately given by the relation:

$$p = 0.4 \ V^2$$

V being the wind velocity at 5 or 6 diameters before the rotor p: the axial pressure on the swept area S. ($p = F_a/S$).

If we write the former expression in the form:

$$p = \frac{1}{2}\rho C_F V^2$$

the value 0.4 corresponds to a thrust coefficient:

$$C_F = 0.64.$$

The rotor being stopped, facing the wind, the axial thrust was equal to 40 % of its value in normal running at equal wind velocity.

The tests performed in the Eiffel Laboratory in Paris on the model of the aerogenerator of Nogent-Le Roi have shown that the coefficient C_F in normal running ($\lambda_0 = 7$) reached, at 20 m/s wind velocity, the value $C_F = 0.418$ and at racing ($\lambda_0 = 14$), the value $C_F = 1.318$. These numbers correspond respectively to values of dynamic pressure given by:

$$p = 0.26 \ V^2 \qquad \text{and} \qquad p = 0.83 \ V^2$$

The experiments which were carried out, have shown, in these two cases, a decrease of the coefficient C_F against the wind velocity.

Note that the application of Euler's theorem with a downwind velocity equal to zero gives, for the dynamic pressure versus V, the value :

$$p = 0.63 \ V^2 \qquad \text{taking for } \rho = 1.26 \ \text{kg/m}^3$$

For the slow wind machines, the axial thrust is more important because of the numerous blades and the existence of a supporting structure. It would be advisable to take :

$$p = V^2$$

(The dynamic air pressure acting on a plate perpendicular to the wind velocity is given as a function of the wind speed by the expression:
$p = 1.25 \ V^2$).

In those relations, p is expressed in N/m^2 and V in m/s.

It should be specified that the calculation of C_F against λ_0 determines the axial thrust for any type of running.

2. THE PITCHING MOMENT

Consequent to the unequal distribution of the wind velocity on the swept area which is characterized by higher speeds in the upper part which can reach two or three times the wind speed in the lower part during gusts, the rotor shaft is subjected to a pitching moment.

This moment may be considerable. For example the pitching moments arising on the French aerogenerator of Nogent-Le Roi at wind speeds upstream of the machine equal to 20 m/s, 40 m/s and 70 m/s, rose respectively :

— in normal running (circumferential speed $U_0 = 75$ m/s) to 8 800, 16 000 and 38 400 daN.m.

Corresponding coefficients C_{PM} $\left(C_{PM} = \dfrac{2MP}{\rho SRV^2} \right)$: 0.032, 0.0149 and 0.0116

— at racing ($U_0 = 270$ m/s) to 29 000, 56 000 and 113 000 daN.m.

Corresponding coefficients C_P = 0.108, 0.0514 and 0.0343.

The values of these moments were determined from the measures carried out in the Eiffel Laboratory in Paris.

3. YAWING MOMENTS

The yaw motor must be capable of driving the turbine around the orientation axis. To determine its power, it is necessary to know the minimum torque necessary to turn the rotor out of the wind direction.

Since each plant constitutes a particular case, we recommend, for powerful installations, to do trial measurements on a model in a wind tunnel.

For the aerogenerator of Nogent-Le Roi, the maximum yawing moments determined from tests done in a wind tunnel at the Eiffel Laboratory in 20 m/s, 40 m/s and 60 m/s winds, reached respectively:

— in normal running: ($U_0 = 75$ m/s): 3 500, 4 750 and 10 250 daN.m.

Corresponding coefficients C_{OM} $\left(C_{OM} = \dfrac{2Mo}{\rho SRV^2} \right)$: 0.013, 0.0044 and 0.003.

— at racing ($U_0 = 270$ m/s): 4 300, 21 800, 45 900 daN.m.

Corresponding coefficients C_{OM}: 0.015, 0.020 and 0.019.

Note that the distance between the yawing axis and the roots of the blades rose for this wind turbine, to 2.25 m.

CHAPTER V

DESCRIPTION AND PERFORMANCES OF VERTICAL AXIS WINDMILLS

Vertical-axis windmills can claim the distinction of being the fore-runners to all wind machines. The successors to the Persian machines of centuries ago have taken many forms. In this chapter the principles of operation and experimental results are described for the most efficient types.

Vertical-axis wind machines can be classified as follows:
— differential drag machines,
— screened machines,
— machines with flapping blades,
— machines with turning blades,
— machines with fixed, moveable or cyclic-pitch blades.

1. DIFFERENTIAL DRAG MACHINES

Because of their symmetry, these machines need no orientation mechanism and their construction is very simple.

The simplest type of differential drag machine is the cup anemometer.

a) PRINCIPLE OF OPERATION

The rotational movement of machines of this type is due to the fact that moving air exerts forces of very different intensity on hollow or unsymmetrical bodies according to their orientation relative to the direction of the wind. If the wind blows on the hollow (concave) side of a hemisphere, the aerodynamic coefficient C which appears in the expression for the driving force :

$$F = \frac{1}{2} \rho S V^2 C$$

is equal to 1.33. It is larger than when the air stream is incident on the convex part of the hemisphere where its value reaches only 0.34. For a half cylinder, these coefficients are 2.3 and 1.2 respectively.

Because of the asymmetry of the constituent parts and the associated differences in air resistance, the action of the wind on the machine as a whole results in a driving couple about its axis so that the rotor is set in motion.

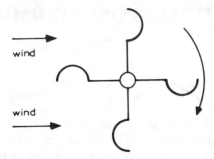

Fig. 93 – *Cupped rotor.*

b) APPROXIMATE THEORY

It is possible to outline an approximate theory for a vertical-axis differential drag windmill.

Let us suppose that the centres of the blades turn with a linear velocity v in a speed V. The aerodynamic forces exerted on the blades are approximately proportional to $(V - v)^2$ when travelling downwind and to $(V + v)^2$ when travelling upwind.

The power developed by the windmill can be expressed as :

$$P = \frac{1}{2} \rho S \left[C_1 (V - v)^2 v - C_2 (V + v)^2 v \right]$$

where C_1 and C_2 are two coefficients presumed to be constant.

This power is optimal when :

$$v = v \text{ opt} = \frac{2 SV - V \sqrt{4S^2 - 3D^2}}{3D}$$

S and D being equal to $(C_1 + C_2)$ and $(C_1 - C_2)$ respectively.

In the particular case when $C_1 = 3 C_2$, we obtain v opt $= V/6$. When $C_2 = 0$, v opt $= V/3$.

For the panemone (fig. 94) $v = \omega R/2$.

In practice, it is established that the output of simple differential-drag machines is optimal for values of the parameter $\lambda_0 = \dfrac{U_0}{V}$ between 0.3 and 0.9, U_0 being the blade tip-speed ωR.

c) PANEMONE

The panemone is one of the oldest differential-impulse machines. The blades are more or less cylindrical.

Conical blades are also used. In his work « L'Homme et le Vent », Auber de la Rue described a vertical-axis windmill he had seen in Canada : It had twelve conical blades and was installed on a reservoir of a farm, where it was used for pumping water. It was a very peculiar type, an elongated version of the sort described above, and operating on the same principle.

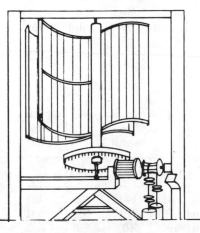

Fig. 94 – *Panemone driving a chaplet pump.*

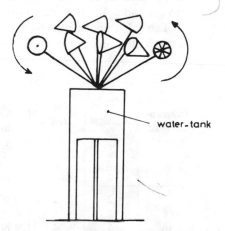

Fig. 95 – *Panemone with conical blades* (According to Aubert de la Rue).

d) LAFOND TURBINE

The Lafond turbine is named after its inventor, who came from Montpellier in France. It is a differential-impulse, cross-flow machine, reminiscent of centrifugal fans and the Banki turbine used in hydraulics.

The aerodynamic force is larger on the blades whose concave face points into the wind than on those who present their convex face.

Moreover, in this machine, the fluid is first deflected by blades moving with the wind and then by the blades moving against it. This results in an additional driving torque.

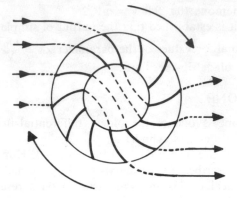

Fig. 96 – *Lafond turbine.*

The Lafond turbine achieves maximum output for values of λ_0 of between 0.4 and 0.4.   It starts to turn in a wind of 2.5 m/s.

Table 14, an extract from « Moteurs à vent » by Champly, shows characteristics and performances of this type of machine.

TABLE 14

Intercepted area (m²)	Diameter (m)	Height (m)	Number & dimensions of blades	Power for various wind speeds (watts)			Turbine mass (kg)
				4 m/s	7 m/s	10 m/s	
4	2	2	24 of 0.5 × 1 m	47	140	400	400
6	2	3	36 of 0.5 × 1 m	70	210	600	600
16	4	4	24 of 1 × 2 m	190	560	1 500	2 000

The output energy of a Lafond turbine is about half that of a horizontal-axis turbine of the same swept area.

e) SAVONIUS ROTOR

This machine was invented by the Finnish engineer Sigurd Savonius in 1924, and was patented in 1929.   Essentially, it consists of two half cylinders whose axes are offset relative to one another.

The original model (see figure 97) was constructed with a ratio of $e/d = 1/3$, e representing the offset of the inside edges and d the diameter of the two half cylinders that make up the rotor.

In addition to the fact that the force exerted on the blades by the wind

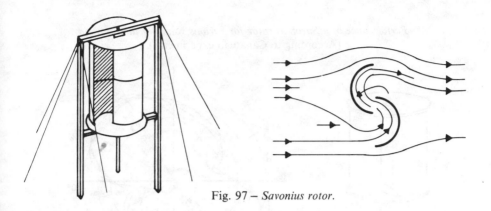

Fig. 97 – *Savonius rotor.*

is different on the convex and concave parts, the rotor is subject to an aerodynamic couple caused by the deflection of the air stream through 180° by the blades (see figure 97).

Experimental work

The Savonius rotor has been the subject of numerous studies notably by the Canadian researchers Newmann and Lek Ah Chai of McGill University in Montreal, who studied the performance of the Savonius rotor for different values of the offset e. The rotors used consisted of half-cylinders 15 inches tall and 6 inches in diameter. Five cases were investigated, each corresponding to a different value of e. (Rotor I, e = 0; II, e = 1; III, e = 1.5; IV, e = 2 and V, e = 2.5 inches).

The experiments measured the starting couple as a function of the position of the rotor with respect to the wind and the delivered power. Figures 98 and 99 reproduce the results and, in particular, specify the starting-torque coefficient as a function of rotor position and the power coefficient as a function of tip-speed ratio.

It will be noted that, for certain rotor positions, the starting torque can be negative. The extent of this zone and the value of the negative couple depends on the size of e. The maximum absolute value of negative torque occurs for e = 0 (rotor I). The corresponding value of C_d is $C_d = 0.40$, and the extent of the zone is 58°. The maximum negative couple is smallest in absolute value for e = 1 inch (rotor II) when $C_d = 0.08$ and the extent of the zone is 18°.

The best results for torque and power are obtained for rotors constructed with a parameter e/d = 1.6 (rotor II). The maximum power coefficient reaches 0.3. For the original Savonius rotor (rotor III e/d = 1.3) the maximum power coefficient is 0.25. The experiments carried out show that the optimal conditions are obtained when the tip-speed ratio $\lambda_0 = \dfrac{\omega R}{V}$ is between 0.9 and 1.

Performance of a Savonius rotor for various values of the ratio e/d
(According to Canadian experiments).

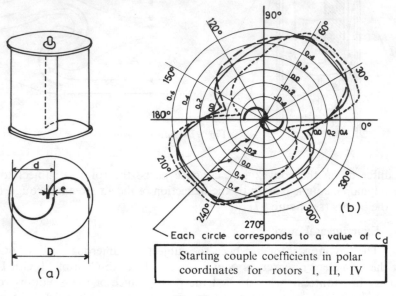

Each circle corresponds to a value of C_d

Starting couple coefficients in polar
coordinates for rotors I, II, IV

Fig. 98

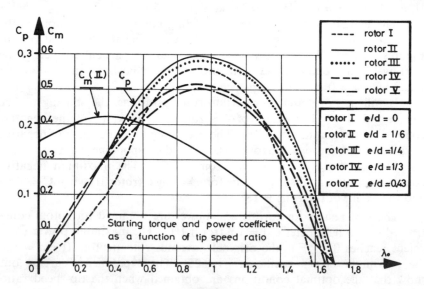

Starting torque and power coefficient
as a function of tip speed ratio

rotor I e/d = 0
rotor II e/d = 1/6
rotor III e/d = 1/4
rotor IV e/d = 1/3
rotor V e/d = 0,43

Fig. 99

Other blade shapes have been tested but up to the present, the performance obtained has always been inferior to that of the rotors described above.

In order to avoid the drawbacks of the negative driving couple for various orientations, some workers had the idea of placing the two rotors one above the other offset by 90º. As far as the power is concerned the performance remains unchanged, but the areas of negative torque disappear.

Available power

The power coefficient C_p is related to power and wind speed by the expression $P = \frac{1}{2} \rho C_p S V^3$. By replacing C_p in this equation by the maximum values obtained in the preceding experiments, 0.30 for rotor II and 0.25 for rotor IV, and by putting $\rho = 1.25$ kg/m^3, the maximum power that can be produced is:

$$P_{max} = 0.18 \, SV^3 \text{ for rotor IV}$$
$$P_{max} = 0.15 \, SV^3 \text{ for rotor IV}$$

S denotes the swept area of the rotor,

$$S = h\,(2d - e) = hD.$$

Due to the increased power coefficient and the enlarged swept area, the maximum power available from rotor II is between 25 and 30 % larger than that obtained from a classical Savonius rotor of the same dimensions. The power is maximised when the tip-speed ratio lies between 0.9 and 1. In general, the power developed by rotor II can be calculated by use of the relation:

$$P = \frac{1}{2} \rho C_p S V_0^3,$$

where $\quad C_p = 0.53 \, (\lambda_0 - 0.2)\,(1.7 - \lambda_0) \quad$ for $0.9 < \lambda_0 < 1.6$

and $\quad C_p = 0.5 \, \lambda_0 - 0.2 \, \lambda_0^3 \quad\quad\quad$ for $\quad 0 < \lambda_0 < 0.9$

Developed torque

The power and torque coefficients are related by the equation $C_p = C_m \lambda_0$. If C_P is known, C_m, and hence the driving torque M are also known:

$$M = \frac{1}{2} \rho C_m R S V^2.$$

In the graph presenting the results of the experiments carried out in Montreal, we have included the curve of C_m for rotor II ($e/d = 1/6$) which is the most efficient of all the models.

Operating machines

We shall limit ourselves to description of some recent installations.

a) CANADIAN MACHINES

The Canadians have explored the possibilities of the Savonius rotor as a means of pumping water for developing countries. The Brace Institute of McGill University in Montreal, has studied a Savonius rotor made up of two 230 litre commercial oil drums of 58 cm diameter (height 1.76 m; swept diameter 0.95 m; e = 19 cm). The rotor studied drove a diaphragm pump. Plans of construction details are given in the publication "How to construct a cheap wind machine for pumping water" published by the research centre.

b) MACHINES OF THE INSTITUT UNIVERSITAIRE DE TECHNOLOGIE DE DAKAR

The first machine that was made, consisted of a stack of four cylinders that had been cut out of a 230 litre commercial oil drum, the upper section was offset by 90º relative to the lower one.

The swept area was 1.8 m². The rotor drove a piston pump of 300 cm² capacity via a chain-reducing gear that lowered the rotational speed by a factor of 7. Table 15 summarises the results obtained by the manufacturers, Bremont and Marie.

TABLE 15

Wind speed m/s	2	3	4	5	6
Power developed by the rotor (W)	4.5	14.5	30.5	42.5	52
Hydraulic power produced (W)	3	9.3	17.3	24.9	18

It will be noted that the machine starts at low wind speeds. One of the advantages of the Savonius rotor is its relatively large starting torque.

Following up on this work, the Institute of Technology of Dakar has undertaken the construction of a series of Savonius rotors of maximum diameter of 2.1 m and height 4 m. These consist of four stacked levels each twisted by 45º with respect to their neighbour. Several systems are under construction.

2. SCREENED MACHINES

In order to eliminate the force of the wind on the blades that are moving upstream, some researchers had the idea of putting a movable screen in front of them. These screens are usually oriented into the cor-

1. Rotating-roofed windmill.

2. Rotating-caged windmill.

3. Neyrpic slow wind turbine coupled to a piston pump.

4. Oasis slow wind machine.

5. The Sept Iles wind generator.

6. MOD.O wind turbine, 100 kW (USA).

7. Balaklava wind machine, 100 kW (USSR).

8. Andreau-Enfield wind turbine 100 kW (G.B.).

10. Vadot Neyrpic wind generator 1 000 kW (France).

9. Vadot Neyrpic wind generator 132 kW (France).

11. Best-Romani wind generator, 800 kW (France)

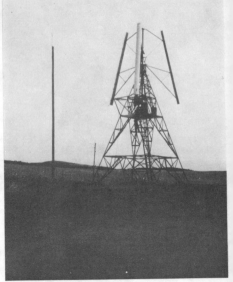

13. Darrieus rotor constructed by J.B. Morel.

12. Darrieus rotor erected on the iles de la Madeleine, 200 kW (Canada).

14. Arnbak windmill (Denmark)

rect position by a downstream vane free to move about the axis of the
machine. Orientation is automatic, see figure 100.

Windmills of this type produce maximal output at values of λ_0 between 0.2 and 0.6.

Ancient Persian windmills

The ancient Persian windmills are related to screened machines but in
their case the screen was stationary. Some were equiped with sails, others
with curtains of reed. Figure 101 shows one of the latter type. Its dimensions were supplied by the National Geographic Society researcher
Hals Wulff.

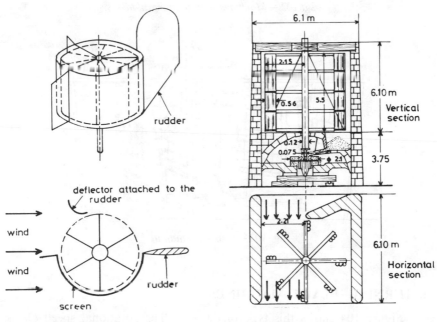

Fig. 100 – *Screened machine.*   Fig. 101 – *Ancient persian windmill.*

3. FLAPPING-BLADE MACHINES

Figures 102 a and b show two types of machines with flapping blades.
The first has a central ring and the second vertical stops.

Maximum output is obtained for $0.2 < \lambda_0 < 0.6$.

There is no need for orientation :

Disadvantages are fragility and noise caused by the impact of the
blades on the stops once per revolution and wear caused by the blade
movement.

Note that the ancient Chinese windmill shown in figure 103 had sails
which operated like the swinging blades.

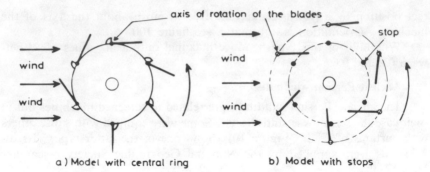

Fig. 102 – *Machines with flapping blades.*

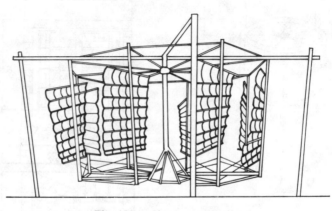

Fig. 103 – *Chinese windmill.*

4. TURNING-BLADE MACHINES

Figure 104 shows this type of device. The rotational speed of the blades is half that of the wheel; their motion is produced by pulleys and belts or by an epicyclic gear. The system is oriented by a vane so that the blade travelling upwind is moved into the eye of the wind. The mechanical losses are large because the complexity of the mechanism requires a large number of pulleys or of cog-wheels and rods.

5. MACHINES WITH FIXED BLADES WITH CYCLIC INCIDENCE
(fig. 105)

Conceived by the French Academician Darrieus, they were patented in 1931. They consist of rigidly connected blades, usually with biconvex profiles that turn about a vertical axis. They have many shapes. The surface described by the blades may be cylindrical, conical, spherical or

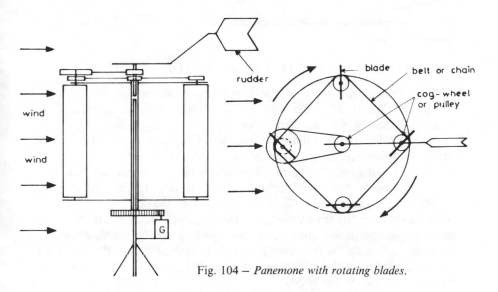

Fig. 104 – *Panemone with rotating blades.*

parabolic. Whatever their appearance, they share a common principle of operation.

a) THEORY

Consider the rotor moving under the action of the wind. Let us examine the aerodynamic forces to which the blades are subject in the various positions that they can occupy (Fig. 105b).

The air speed $\vec{W}$ relative to the blades is related to the wind speed $\vec{V}$ and the peripheral speed $\vec{U}$ by the equation: $\vec{V} = \vec{U} + \vec{W}$ which can be rewritten: $\vec{W} = \vec{V} - \vec{U}$. If the vectors $\vec{V}$ and $\vec{U}$ are known, it is possible to determine the vector $\vec{W}$ and hence the aerodynamic forces to which the blades are subjected. If a constant wind speed and direction accross the rotor is assumed, this calculation does not present much difficulty.

Study of the velocity triangle for different positions of the blades shows that the forces create a driving torque in all positions, except when the plane of symmetry of the profile of a blade element is parallel or nearly parallel to the wing direction.

The variable angle of attack of the relative wind on the profiles never exceeds the limiting value $i_{max} = \sin^{-1} (V/U)$. These angles are relatively acute if the peripheral speed U of the rotor is large compared to the wind speed (V). This fact makes an acceptable aerodynamic output feasible.

On the other hand, if the windmill is stationary, the relative speed $\vec{W}$ is the same as the wind speed $\vec{V}$, The angles of incidence in the system are therefore much larger, and for certain positions, stall occurs. The starting couple is very small and in practice Darrieus windmills have to be started externally.

b) STUDIES OF THE DARRIEUS ROTOR

The theoretical study of the Darrieus rotor with parabolic blades was undertaken by R. J. Templin, Head of the Canadian National Aeronautical Research Council laboratory.

In the following, this theory as applied to Darrieus rotors with vertical, straight sloping, spherical and parabolic blades is discussed.

We should immediately recall an important aerodynamic result:

For a blade positioned obliquely with respect to the leading edge, the lift is the same as if the blade were subject to the component $V_0 \cos \Phi$ perpendicular to the leading edge. The angle of incidence is defined in terms of the component directions: one lying along the profile chord and the other following the direction mutually perpendicular to the chord and the leading edge. The component $V_0 \sin \Phi$ parallel to the leading edge has no effect. (see figure 105 d).

Let us now imagine a fixed set of axes Oxyz and a Darrieus rotor rotating about a vertical axis which is coincident with Oz. The absolute speed of the wind across the rotor is V and its direction is parallel to Ox (Fig. 105 c).

Consider a blade element centred at M of chord 1 and length ds. Let r be the distance of this element from the axis of rotation and θ be the angle between the plane Oyz and the vertical plane containing the axis of rotation and the perpendicular at M to the chord of the blade element considered. (We shall assume that this plane is unique for all the elements of the same blade). Let δ be the angle between the normal to the blade element and the horizontal plane.

For a parabolic rotor of height 2H centred at 0:

$$\frac{r}{R} = 1 - \frac{z^2}{H^2}$$

$$\delta = \tan^{-1}(2zR/H^2)$$

For a cylindrical rotor with rectangular vertical blades:

$$r = R \text{ and } \delta = 0.$$

For a truncated conical rotor:

$$r = R_0 - (R_0 - R_1)\, z/H$$
$$\delta = \tan^{-1}((R_0 - R_1)/H)$$

Let us calculate the components of the relative speed $\vec{W}$ in the direction of the chord of the blade element and mutually perpendicular to this chord and the leading edge. For this calculation, let us adopt the following auxiliary axes: (see figs. 105c to 105e):

Mz' vertical, positive upwards, Mt chordwise direction, positive moving from leading to trailing edge, Mr horizontal, perpendicular to the chord.

The relative velocity $\vec{W}$ is related to the tangential speed $U = \omega r$ and to the absolute wind velocity $\vec{V}$ by the equation:

$$\vec{W} = \vec{V} - \vec{U}.$$

The vector $\vec{W}$ is the sum of two horizontal vectors and so is itself horizontal. The components of $\vec{W}$ in the directions defined above are:

$$W_r = V \sin \theta, \quad W_t = U + V \cos \theta, \quad W_{z'} = 0.$$

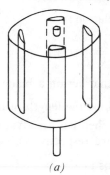

(a)

Cylindrical Darrieus rotor

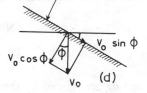

(b)

$\vec{W} = \vec{V} - \vec{U}$

Wind action

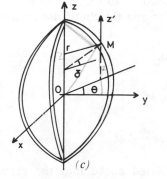

(c)

Parabolic D. rotor

(d)

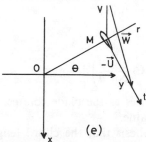

(e)

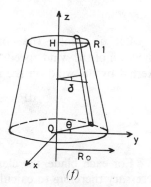

(f)

Truncated conical D. rotor

Fig. 105 – Various Darrieus rotors.

The directional cosines of the normal to the blade element are, in the same system, cos θ, 0 and sin θ.

It follows from this, that the component of $\vec{W}$ perpendicular to the blade element, is V sin θ cos δ. In other words:

$$W_t = U + V \cos \theta = r\omega + V \cos \theta,$$

The speed W used to determine the forces to which the blade is subjected, is such that:

$$W^2 = (r\omega + V \cos \theta)^2 + V^2 \sin^2 \theta \cos^2 \delta,$$

The angle of incidence is defined by the relation:

$$\tan i = \frac{V \sin \theta \cos \delta}{r\omega + V \cos \theta}$$

Let us evaluate the components of force that act on a blade element. Let the dynamic pressure be $q = \frac{1}{2} \rho W^2$ and l, be the length of the chord of the profile considered. C_n and C_t are the Lilienthal aerodynamic coefficients (parallel and perpendicular to the chord respectively),

$$C_t = C_l \sin i - C_d \cos i$$
$$C_n = C_l \cos i + C_d \sin i$$

ds is the length of blade element projected on to the leading edge.

The components of the aerodynamic force in the chordwise direction and normal to the blade element are:

$$dN = C_n ql \frac{dz}{\cos \delta}$$
$$dT = C_t ql \frac{dz}{\cos \delta}$$

ds and dz are related by the equation

$$dz = ds \cos \delta$$

Let us resolve the components calculated above into the general direction of the wind speed V and let us calculate the resultant of the force exerted by the wind on the rotor in that direction:

$$dF = dN \cos \delta \sin \theta - dT \cos \theta$$

$$= ql\left(C_n \sin \theta - C_t \frac{\cos \theta}{\cos \delta}\right) dz$$

For each blade, the elemental force varies as the blade rotates. It is necessary therefore to calculate the mean value.

Given these conditions and the hypothesis that the chord length of the windmill blade is constant, the force in the direction of the wind exerted on the rotor as a whole is given by the equation:

$$F = \frac{pl}{2\pi} \int_{-H}^{+H} \int_{0}^{2\pi} q\left(C_n \sin\theta - C_t \frac{\cos\theta}{\cos\delta}\right) d\theta dz.$$

Derivation of expressions for torque and power

The moment of the aerodynamic forces exerted on the blade element about the axis of rotation is equal to:

$$dM = \frac{C_t ql}{\cos\delta} r dz$$

For the complete rotor, the moment is:

$$M = \frac{pl}{2\pi} \int_{-H}^{+H} \int_{0}^{2\pi} \frac{C_t qr}{\cos\delta} d\theta dz$$

Hence the power becomes:

$$P = M\omega = \frac{pl}{2\pi} \int_{-H}^{+H} \int_{0}^{2\pi} \frac{C_t qr\omega}{\cos\delta} d\theta dz$$

The sequence of calculations does not follow that of R. J. Templin, but the technique applied, is based on the same principle and the results are identical.

The expression above, taken alone does not allow the determination of the performance of a rotor in a wind speed V_1. To obtain this, it is necessary to use the Betz theory.

We have seen that the force exerted on a horizontal axis wind turbine is given by the relation:

$$F = \rho SV(V_1 - V_2)$$

On taking $V_2 = kV_1$, the relation giving the speed through the rotor can be written as:

$$V = \frac{1}{2}(V_1 - V_2) = V_1 \frac{(1+k)}{2}$$

Thus the expression of F becomes:

$$F = \frac{1}{2}\rho S(V_1^2 - V_2^2) = \frac{1}{2}\rho SV_1^2(1-k^2) = 2\rho SV^2 \frac{1-k}{1+k}$$

Let us assume that the Betz theory is valid for vertical-axis wind rotors. By equating the previous expression to the value F obtained for the vertical axis wind turbine, it follows that:

$$2\rho SV^2\left(\frac{1-k}{1+k}\right) = \frac{bl}{2\pi} \int_{-H}^{+H} \int_{0}^{2\pi} q\left(C_n \sin\theta - C_t \frac{\cos\theta}{\cos\delta}\right) d\theta dz$$

Replacing q by its value $\frac{1}{2}\rho W_u^2$, we obtain:

$$G = \frac{1-k}{1+k} = \frac{bl}{8\pi S} \int_0^\pi \frac{W_u^2}{V^2} \left(C_n \sin\theta - C_t \frac{\cos\theta}{\cos\delta} \right) d\theta\,dz$$

where:

$$\frac{W_u^2}{V^2} = \left(\frac{r\omega}{V} + \cos\theta \right)^2 + \sin^2\theta\,\cos^2\delta$$

Since
$$\frac{\omega r}{V} = \frac{r}{R}\frac{\omega R}{V}, \text{ we can write:}$$

$$\frac{W_u^2}{V^2} \left(\frac{\omega R}{V}\frac{r}{R}\cos\theta \right)^2 + \sin^2\theta\,\cos^2\delta$$

Likewise:

$$\tan i = \frac{\sin\theta\,\cos\delta}{\dfrac{\omega R}{V}\dfrac{r}{R} + \cos\theta}$$

It is therefore possible given the ratio $\omega R/V$ to calculate G and hence k:

$$k = \frac{1-G}{1+G}$$

Knowing k allows calculation of the speed ratio λ_0:

$$\lambda_0 = \frac{\omega R}{V_1} = \frac{\omega R}{V}\left(\frac{1+k}{2} \right) = \frac{\omega R}{V(1+G)}$$

Power and torque coefficients

The power coefficient is defined as:

$$C_p = \frac{2P}{\rho S V_1^3} = \frac{bl}{2\pi S} \int_{-H}^{+H} \int_0^{2\pi} C_t \frac{W_u^2}{V_1^3} \frac{\omega r}{\cos\delta}\,dz\,d\theta$$

We can write:

$$\frac{W_u^2}{V_1^3}\omega r = \frac{W_u^2}{8V^2}\frac{\omega R}{V}\frac{r}{R}(1+k)^3$$

By using this equation, C_p may be calculated for various values of $\omega R/V$. To each value of $\omega R/V$, corresponds a value of λ_0. Therefore plotting the C_p curve as a function of the speed ratio λ_0 does not present any difficulty.

The moment coefficient is related to the power coefficient by the equation:

$$C_p = C_m\,\lambda_0$$

Thus, knowledge of the power coefficient automatically includes the moment coefficient. This will be defined for the vertical axis rotor by the expression :

$$C_m = \frac{C_p}{\lambda_0} = \frac{2P}{\rho SV_1^3} \frac{V_1}{\omega R} = \frac{2M}{\rho SRV_1^2}$$

Remarks

1. Note that in the quoted expressions, the shape of the rotor is defined by the limits of the integrals and the value of r as a function of z and cos δ. Consequently, a general calculation scheme could be devised such that it was applicable to various shapes of windmill: cylindrical, truncated conical, spherical and parabolic without much modification, except perhaps for the z and cos δ variations. It would only be necessary to alter the geometric dimensions of the windmill and the data describing the chosen profile which would, of course, be inserted into the calculation scheme.

2. Furthermore, it should be noted that in the theory described, it is assumed that the lift and drag coefficients are the same in steady and unsteady flow for a given incidence. This is a substantial assumption and requires verification. Templin states that results thus obtained agree well with experimental data.

3. Consideration of symmetry often allows considerable reduction in the volume of calculation. For this reason, for rectangular, spherical or parabolic windmills, the calculations need only be performed for the part of the windmill above the equatorial plane. This alters neither the value of k nor the power coefficient on condition that the value of S used correspond to the swept area of that part of the windmill.

In the same way, if it is assumed that wind speed V is the same for the upwind and downwind blade and that the profiles used are symmetrical, the resulting symmetry in the velocity triangles allows the calculation to be limited to the interval :

$$0 < \theta < 180^\circ$$

Under these conditions, it is necessary to double the integrals used to calculate k and C .

4. Tests and mathematical models carried out by J. Templin, in Ottawa, on parabolic Darrieus rotors with two and three blades show that their maximal power is approximately given by the following relationship :

$$P = 0.25 \, SV^3$$

This power corresponds to a speed ratio λ_0 related to the chord l of the blades by the relation :

$$\lambda_0^2 = \frac{5R}{bl}$$

c) PRACTICAL SYSTEMS

The most notable installations of Darrieus windmills have been carried out in France, the USA and Canada.

In France, the research was done jointly by Electricité de France and J. B. Morel of Grenoble. In the USA, the Darrieus rotors were studied by NASA and in Canada, the installations were the work of the National Canadian Research Council.

1°) French installations : J. B. Morel windmills

J. B. Morel has constructed three windmills on behalf of Electricité de France :

— an experimental windmill of 1.75 m span with three blades which developed 300 W with an efficiency of 23 % at a wind speed of 10 m/s in a wind tunnel in St Cyr,

— one windmill of 5 kW and one of 7 kW.

The blades of these last two windmills were slightly inclined to the vertical and so formed a truncated cone rotating about a vertical axis. This shape allows the use of bracing. The results obtained with these two machines are presented in table 16.

TABLE 16

Rated power kW	5	7
Diameter of rotor (m)		
at Top	2	4.2
at Bottom	8.5	8.5
Height of rotor (m)	5.7	5.7
Rotational speed (r.p.m.)	50	
Wind speed (m/s)	10	
Efficiency		0.23

Tests made on the 5 kW windmill showed that the efficiency wind speed curve had a broader plateau of maximum efficiency than the 7 kW machine whose shape is closer to that of a cylinder. Several windmill projects of this type were studied by Electricité de France and J. B. Morel.

Table 17 shows the main characteristics of a proposed system taken from an article by R. Bonnefille which appeared in « La Houille Blanche » in January 1975.

TABLE 17

Rated power kW	90	500	1 200
Diameter of the rotor (m)			
Top	3.9	12	18
Bottom	20	40	60
Height of rotor (m)	14	36	54
Rotational speed (r.p.m.)	50	14.3	14.3
Wind speed (m/s)	15	15	15

These plans, based on an efficiency of 25 %, were never to be carried out.

2°) American studies

NASA studied a Darrieus rotor consisting of three thin, curved plates with plastic aerofoil sections whose ends were fixed to the axis of rotation. The rotor produced 1 kW in a 6.66 m/s wind. It was self-starting in a 5.2 m/s wind, but only rotated at a speed of 13 r.p.m. If the rotor is spun up to 65 r.p.m., the speed increases until it stabilises at 213 r.p.m.

3°) Canadian studies and installations

The most important installation and research in the field of vertical axis turbines has been done by the Canadians, notably by the researchers of the National Research Council of Canada, Peter South and Raj Rangi.

The studies were carried out on Darrieus rotors with two or three blades made from thin curved plates. Two two-bladed models are currently commercially available through Dominion Aluminium Fabricating (fig. 106). One model 4.6 m in diameter can produce a maximum of 4 kW at 115 Volts; another model, 6.1 kW at 115 Volts.

The dimensions are shown in table 18.

TABLE 18

Rotor diameter	A	B	C	D
4.6 m	4.6	5.5	2.4	11.7
6.0 m	600	9.1	2.4	13.7

The machines are built to withstand winds of 45 m/s with gusts to 60 m/s. They start to produce energy when the wind speed exceeds 3.6 m/s. When the wind is not blowing sufficiently hard, the windmill does not rotate.

The rotor is started automatically by a battery-powered motor. When the speed is high enough an alternator takes over from the motor and charges the batteries. In order to provide protection against winds which are too strong, brake flaps are deployed when the rotational speed exceeds a given value.

Darrieus rotor of the Dominion
(Aluminium fabricating).

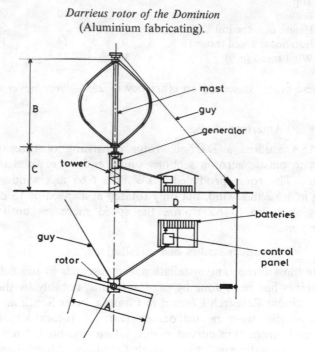

Fig. 106a

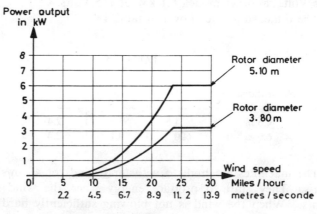

Fig. 106b

The mean monthly output in kWh is given in table 19 as a function of wind speed.

TABLE 19

Rotor diameter (m)	Voltage Volts	Wind speed m/s				
		4	5	6	7	8
4.6	24	110	200	310	500	670
4.6	110	110	200	390	640	990
6.1	110	210	420	745	1 040	1 860

The overall efficiency of the machines relative to the Betz limit is 68 %, which is excellent for this type of machine, bearing in mind the simplicity of construction and the capital cost. The Canadian installations do not stop there.

The National Research Council has constructed a two-bladed vertical axis windmill at Iles de la Madeleine in the Golfe du St Laurent. The blades form a ring of 37 m diameter and 24 m in height. The machine supplies 200 kW to the local grid energy which would otherwise come from imported oil. Two flat plates are fixed to each windmill blade in the equatorial plane and fastened along the two axes of the profile. Their function is to provide aerodynamic braking as soon as the rotational speed of the machine exceeds a certain value. These plates are called spoilers and under normal conditions are held flush with the blades by springs. Their centres of mass do not lie on the pivot axis and so they move away from their normal position under the action of centrifugal force when the limiting speed is reached. The machine is also equipped with a mechanical brake.

6. COMPOSITE SAVONIUS-DARRIEUS SYSTEMS

An inconvenient characteristic of the Darrieus rotor is that it requires an external starting torque in order to reach its operating conditions. The Canadians have overcome this problem by equipping the rotor with a motor that can also serve as a generator.

It is equally possible to start the Darrieus rotor with the aid of a Savonius rotor. Figs. 107 (a) and (b) show two systems which were experimented with in the USA and USSR.

Due to the large starting torque produced by the Savonius rotor, the starting of the whole system is enhanced but the power produced under normal operating conditions is essentially due to the Darrieus rotor.

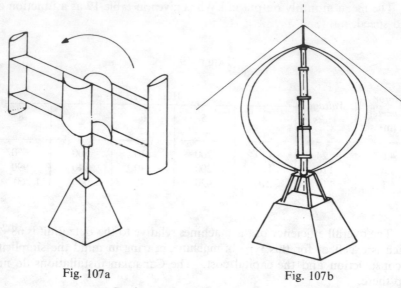

Fig. 107a   Fig. 107b

Combined rotors (Savonius-Darrieus).

7. LOW SPECIFIC SPEED ROTOR DESIGNED BY PROFESSOR NGUYEN VINH

A rotor consisting of three blades made up of a half-cylinder combined with a flat plate (see fig. 108) was studied at the École Polytechnique de Thiès (Senegal) by Professor Nguyen Vinh and his collaborator Mr. Houmaire. For a rotor, in which the diameter d of the semi-circles and the length l of the plane part placed behind the semi-circles are approximately equal to one third of the diameter D, the coefficient C passes through a maximum of about 0.3 for a tip-speed ratio of $\lambda_0 = 0.7$. Furthermore, the starting torque is large because of the peculiar shape of the blades.

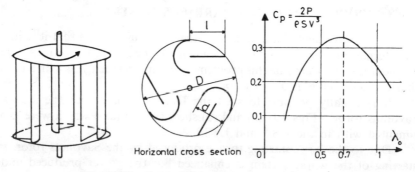

Horizontal cross section

Fig. 108a – *Rotor due to Professor Nguyen Vinh.*

The performance of the rotor studied was comparable to that of the Savonius rotor, but its construction requires considerably less material for the same power and hence the rotor is lighter.

8. ANOTHER FIXED-BLADE MACHINE WITH CYCLIC INCIDENCE: THE ARNBAK WINDMILL

This windmill designed by the Danish engineer Lars Arnbak is reminiscent in shape, of the Savonius. In fact its mode of operation is based on the same principles as the Darrieus rotor. It consists of two indentical cylindrical parts; each has an elliptical section and is offset by 90º with respect to the other in order to make the torque more uniform and to facilitate starting.

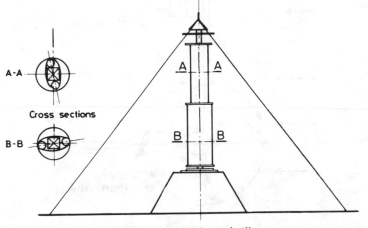

Fig. 108 – *Arnbak windmill.*

9. WINDMILLS WITH MOVING BLADES

a) CYCLOGYRO

The cyclogyro or gyromill is not just another Darrieus rotor with moving blades. The variation of blade inclination is achieved by guide rods or cams. The principle of the system is illustrated in figure 109. Trials have shown that the cyclogyro has better efficiency than the classical Darrieus rotor at low speeds. The machine has the advantage of being self-starting.

The model sold by the Pinson Energy Corporation of Marston Mills, Massachussetts, consists of three eight-foot blades. The diameter of the rotor that develops 2 kW in 10 m/s wind is 3.65 m.

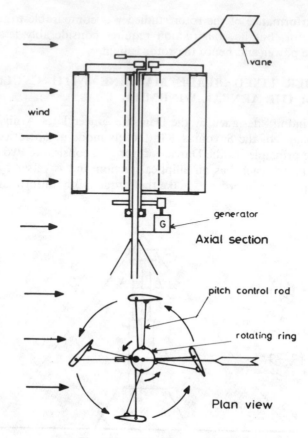

Fig. 109 – *Giromill or cyclogiro.*

b) DARRIEUS ROTOR DESIGNED BY DE LAGARDE (UNI-VERSITY OF MONTPELLIER) AND EVANS (UNIVERSITY OF ST ANDREWS)

These are Darrieus rotors with moving blades. The centres of gravity of the blades are eccentric relative to the axis of rotation. The resultant of the centrifugal forces on the centres of gravity makes the blades turn through a certain angle. This results in the machine being self-regulating.

In the de Lagarde machine, this angle is limited by two stops and an elastic tie attached to the trailing edge. In the Evans machine this is done by a runner and a spring which applies a force to the leading edge (figs. 110 and 111). In comparison with the classical Darrieus rotors, the two machines have the advantage of being self-starting.

Darrieus rotors with oscillating blades

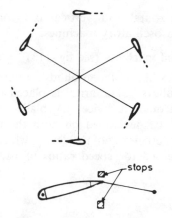

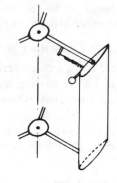

Fig. 110 – *De Lagarde design.*   Fig. 111 – *Evans design.*

c) VARIABLE-GEOMETRY DARRIEUS ROTOR

In trying to limit the speed of rotation, P. J. Musgrove and I. D. Mays of Reading University in the UK have studied Darrieus rotors with rectangular blades and variable geometry (fig. 112).   Under the action of centrifugal force, which is exerted unequally on the parts of the blades on either side of the secondary axes of rotation, the truncated cone described by the blades changes shape and causes a reduction in the intercepting area and the speed of rotation.   The machines of this type are self-regulating.

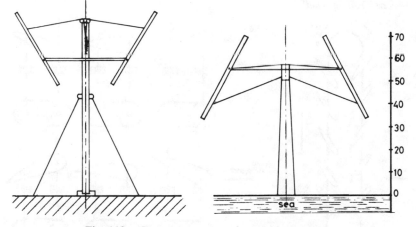

Fig. 112 – *Darrieus rotors with variable geometry.*

a) experimental model   b) high power prototype
   (University of READING)   diameter ⌀ = 75 m, P = 2 500 kW

10. OTHER TYPES OF MACHINES

In addition to the classical horizontal and vertical-axis windmills there are other types: translatory and oscillatory machines.

1°) **Translatory or rolling-band machines** (see fig. 113)

These consist of an array of blades like venetian blinds mounted on a system of bands passing over two pulleys. The array is placed perpendicular to the wind which flows through it twice. A blade-reversing mechanism allows a driving force to be developed on both the upwind and downwind blades. The maximum efficiency of the device, which needs to be oriented into the wind, is obtained for speed ratios of between 2 and 3.

2°) **Oscillatory machines**

a) OSCILLATING AEROFOIL MACHINES TRANSFORMING RECIPROCATING MOTION INTO ROTATIONAL MOTION (fig. 114)

The essential element is an aerofoil section mounted on a system of articulated rods attached to pulley wheels. The mechanical linkage is such that the angle of incidence changes continuously. When the leading edge of the aerofoil is in the upper position, the trailing edge is at its lowest and vice-versa. In order to avoid dead points at the top and bottom

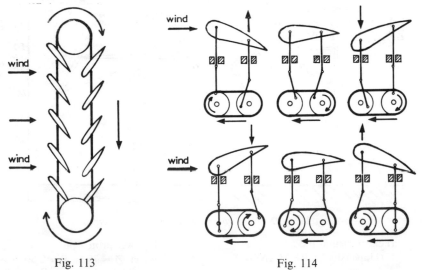

Wind machines using moving shutters known as " rolling band"

Wind machine using an oscillating blade

Fig. 113   Fig. 114

positions, the top dead centres of the points of attachment of the rods to the pulleys are separated by angles of between 45° and 60°.

The links transform the reciprocating motion into rotational motion.

b) MACHINE WITH SWINGING BLADES THAT TRANSFORM OSCILLATORY MOTION INTO TRANSLATORY MOVEMENT (fig. 115)

This machine was conceived by Peter Bade. It principally consists of a blade free to rotate about an axis fixed on an oscillating arm. The oscillating movement of the blade, whose amplitude is limited by solid stops on a cable, is used to drive a piston pump directly.

Pump using a swinging blade

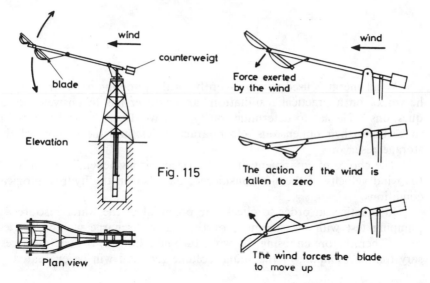

Fig. 115

(Source : Revue « Écologie »)

CHAPTER VI

USE OF WIND ENERGY
FOR WATER PUMPING

Wind energy is used very frequently for the pumping of water. When he works on a practical installation, an engineer has to answer several questions. He has to determine the type of wind turbine and the pump to be used, their dimensions and characteristics, and the volume of the storage reservoir.

The choice of equipment and its characteristics depends not only on the wind velocity at the site considered, but also on the hydrogeological conditions.

In practice, in order to check the potential of the water resource, a pumping test with measurement of the water discharge will be carried out. Then, before choosing the wind turbine diameter, the energy necessary for the pumping of the water volume required will be evaluated.

A. Preliminary Studies

1. WATER SUPPLY STUDY

This study is indispensable for a well or a borehole. It can be conducted following classical methods : Porchet's method or a method based on well equilibrium.

The results obtained are depicted in figure 116. The water quantity removed from the well Q is plotted on the abscissa, and the drawdown of the water level in the well y is plotted on the ordinate.

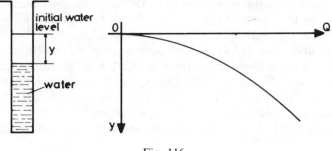

Fig. 116

2. WATER DEMAND STUDY.   ASSESSING THE ENERGY NEEDED FOR PUMPING

The water demand is not necessarily constant during the whole year but may vary.   The variation depends on whether the water is used for drinking or irrigation (fig. 117).

It has to be remembered that the demand can be satisfied only if it does not exceed the available water supply.   We shall assume that the well is capable of supplying all the water needed.

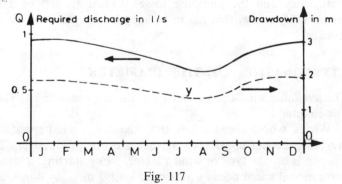

Fig. 117

As the results of the water pumping tests are now available, the water level in the well corresponding to any demand can be obtained. The variation of water level can be plotted on the same graph as the variation of water demand.

If the elevation of the storage reservoir, which has to be filled with water, is known, it is simple to calculate the pumping losses when the diameter of the pipe between the pump and reservoir is known.   The calculation may be based, for example, on an average speed in the pipe from 0.8 to 1 m/s (approximate economic water speeds).

If h is the difference in elevation between the initial level of the water table in the well, y the drawdown and ΔH the pumping losses, the power expended by the pump to supply the demand Q is given by: $P_u = \overline{\omega} Q Y$ where $Y = h + y + \Delta H$ is the manometric head.

As Q and y are known, Y and consequently P_u can be calculated at any time. A curve showing the variation of P_u as a function of time can be drawn without difficulty on the same graph as that showing y and Y. The area, between the curve of P_u and the time axis, gives the energy supplied by the pump for raising the water into the reservoir.

The energy E_u required for the pumping of water will be calculated for each month.

Actually, the amounts of energy that will have to be supplied by the wind turbine are larger than the values determined so far because of the irregularity of the wind and the fact that the pump is not perfectly efficient. The amount of water pumped at any instant is not equal to that of the consumption. At times, it is zero; at other times, it is very large. During periods of high wind, the pumping rate increases and the water level drops at a greater than normal rate. Consequently, the pumping losses and the monthly energy expenditures increase above those calculated for average conditions.

In practice, the monthly energy requirement E_u, calculated above, will be multiplied by a coefficient of 2 to 3 (at least 2) to account for the pump efficiency and the pumping losses. Then $E_p = 2$ or $3 E_u$ is the effective monthly energy that has to be supplied for pumping during sucessive months.

3. DETERMINATION OF THE DIAMETER

The available wind energy will be evaluated according to the method given in chapter I.

— Winds, whose speed is less than the cut-in wind speed, V_m of the wind turbine, are not considered as they are not productive. The value of V_m depends on the type of wind turbine. For starting, a wind turbine of low rotational speed needs winds of at least 3 m/s, say 4 m/s, if it drives a piston pump. High-speed wind turbines require winds of 5 m/s to 5.5 m/s.

— Strong winds exceeding V_M, which endanger the installation, and during which the wind turbine is stopped, are also discarded.

— In winds which exceed the rated wind speed V_N, and for which regulating devices come into play, the wind turbine does not produce more energy than in winds of speed V_N. These higher wind speeds have to be counted, therefore, as having value of V_N.

It should be noted on this subject, that the values of V_M and V_N are, in general, much lower for low-speed wind turbines than for high-speed wind turbines.

Then, from the available wind speed data, the energy e_c, which can be captured per m² of rotor area, for the pumping of water, will be evaluated for each month.

When the type of machine will be chosen, the diameter D of the wind turbine will be calculated, to assure that the desired quantities of water can be supplied for each of the successive months.

The maximum diameter obtained will be selected as the effective diameter to give to the wind rotor. Should a smaller diameter machine be chosen, it has to be understood that the water supply is likely to be insufficient from time to time.

In practice, two types of installation are available: low-speed wind turbines operating piston pumps and high speed machines driving helical or turbine pumps. Other devices can also be envisaged, but the above types are the most common ones.

B. Water Pumping by Low-Speed Windmills and Piston Pumps

1. INTRODUCTION

Among the different types of pump that can be used with multibladed windmills, the piston pump is the most suitable because it runs well at low speeds. It accommodates perfectly the slow-speed rotation of the multibladed windmill. A single-acting pump is used most of the time. Such an installation is shown in figure 118. Note the foot valve and the valve in the piston which opens during the downward stroke of the piston. It should also be noted that, with such a pump, the piston rod always operates in tension, during the up-stroke as well as during the down-stroke. Therefore, it does not need to be guided to avoid buckling.

2. TRANSMISSION OF MOTION FROM WINDMILL TO PUMP

Various mechanisms are used to transform the rotational motion of the windmill into a reciprocating motion that can be utilized by the pump.

One of these mechanisms is shown in figure 118. A pinion fastened to the windmill axis drives a large gear which is attached to the piston rod by means of a connecting link, the end of the piston rod is constrained by a secondary link to move on a circular arc. As the depth of the arc is small, the motion of the piston rod can be considered to be linear.

A similar mechanism is shown in figure 119a and a mechanism based on an eccentric is shown in figure 119b.

In the latter mechanism, a cam is fixed to the windmill axis. There is an up-and a down-stroke during each revolution of the windmill. Friction is reduced by means of rollers.

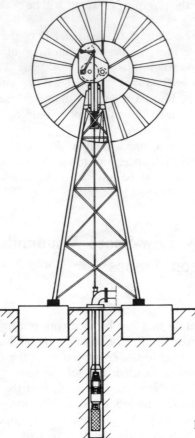

Fig. 118 – *Slow wind turbine driving a piston pump,*

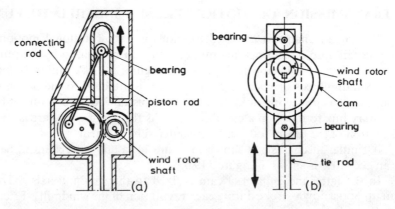

Fig. 119

3. INFLUENCE OF THE MECHANISM ON THE STARTING SPEED

Let us evaluate the force which acts on the piston rod during the up-stroke of the piston.

If   P   is the weight of the moving parts,
   H, the total static head,
   S, the cross sectional area of the piston,
   $\overline{\omega}$, the specific weight of the water,

then the vertical force F which has to be overcome for raising the water is :

$$F = P + \overline{\omega} SH$$

If the radius of the circle described by the end of the connecting rod is "a", the moment C to be overcome is :

$$C = aF = a(P + \overline{\omega} SH)$$

Under these circumstances, if n_2 represents the number of teeth of the connecting rod driving gear, n_1 the number of teeth of the pinion, and k the ratio n_1/n_2, the maximum torque C_1 transmitted to the rotor axis is given by :

$$C_1 = ka(P + \overline{\omega} SH)$$

The lower the maximum torque to be overcome, the easier will the windmill start. If the necessary torque is high, a faster wind speed is required to start the windmill. The operating time of the machine is consequently reduced.

It is therefore desirable to reduce the torque C_1. To reduce it, and hence to make the starting easier, various actions are possible :
   — changing the radius "a",
   — changing the speed-up ratio $k = n_1/n_2$,
   — smoothing the driving torque.

a) CHANGING THE RADIUS "a"

The expression for C_1 shows that a reduction of the operating torque can be obtained by reducing "a". But if "a" is reduced, the piston stroke diminishes and consequently, the volume of water q delivered with each stroke of the piston decreases. This is, in effect linked to "a" through the relation :

$$q = 2aS.$$

If we want to maintain the delivery at its initial value, it is necessary to change the piston area according to the relation :

$$S = \frac{q}{2a} \text{ with q being kept constant.}$$

Replacing S, by the above value, in the expression for the torque C_1 which is transmitted to the windmill axis, it follows that :

$$C_1 = k\left(aP + \bar{\omega}\,\frac{qH}{2}\right)$$

This equation shows that, at constant delivery of water, the operating torque decreases with decreasing a.   Thus for reducing the starting torque of a windmill, " a " should be reduced and the piston area should be increased.   For wells of small depth (30 to 40 m), this can be done without difficulty.   For deep wells, this change becomes uneconomical as the cost of sinking the well increases rapidly with an increase in diameter.

In general, the manufacturers supply each windmill with two strokes. They suggest a small stroke with large piston diameter for shallower wells, and a long stroke with small piston diameter for deep wells.

b) CHANGING THE SPEED-UP RATIO : $k = n_1/n_2$

The expression for the moment :

$$C_1 = k\left(aP + \bar{\omega}\,\frac{qH}{2}\right)$$

shows that a reduction of k leads to a reduction of C_1.   But if we reduce the value of k, the number of pump strokes per second N decreases.   Consequently, less water will be supplied because Q and q are related by the expression :

$$Q = kNq$$

In order to maintain a constant output volume Q, the value of q has to be increased.   For a given value of a, we can increase S.   Under these conditions, we obtain :

$$C_1 = kaP + \bar{\omega}\,\frac{Q}{N}\cdot\frac{H}{2}$$

This relation shows that a reduction of the ratio k has the same effect on the starting velocity as a reduction in radius a.   The weight of the moving parts is thus reduced.

In practice, the ratio 1/k is never greater than five.

4. SMOOTHING THE PUMPING TORQUE

In spite of the improvements which can be obtained through changing the radius a and the gear ratio, the torque applied by the pump on the windmill axis is a fluctuating torque. The energy furnished by the windmill is absorbed mainly in raising the water and the piston when this latter moves up. The piston goes down under its own weight.

This has a direct adverse effect on the starting speed. In fact, a much larger torque than the average torque, and consequently a much higher wind speed is required to start the machine. In order to reduce the necessary starting wind speed, and thereby increase the yearly operating time of the machine, this fluctuation should be smoothed out as much as possible.

A double-effect pump might be used which would halve the value of q but, in that case, the piston rod is subject to alternate tension and compression.

In fact, this solution is rarely employed. It is better to use single acting pumps and smooth the torque by other means: counterweights or springs. A system which utilizes a spring, is particularly simple. The slow-running windmill shown in figure 118 is equipped with such a device.

The spring acts on the piston rod, imparting an upward force during the rise of the piston. The intensity of the vertical force F, which is transmitted to the arm during this phase of the motion, is:

$$F = P + \overline{\omega}SH - F_1$$

The maximum torque applied to the windmill axis during the same period becomes:

$$C_1 = ka(P + \overline{\omega}SH - F_1)$$

The force F_1 continues to act during the downstroke of the piston. But is has a resisting effect. This may be higher than the effect of the weight of the moving parts. In that case, the torque applied to the windmill axis during the downstroke of the piston is also an opposing torque. Its maximum value rises to:

$$C_2 = ka(F_1 - P)$$

Smoothest possible windmill rotation is obtained when the condition $C_1 = C_2$ is satisfied, i.e. when:

$$P + \overline{\omega}SH - F_1 = F_1 - P$$

or

$$F_1 = P + \frac{\overline{\omega}HS}{2}$$

The value of the maximum torque then becomes:

$$C = ka\frac{\overline{\omega}SH}{2} = \overline{\omega}kq\frac{H}{4} = \frac{\overline{\omega}QH}{4N}$$

To smooth the torque, it is possible to use a counterweight and a lever which supply a force F_1 of constant magnitude or a spring.   In that case F_1 is equal to kx, x being the elongation (fig. 120a and b).

It should be noted that with the latter device, one should preferably select a long spring so that the force F_1 does not vary greatly thoughout the full stroke of the pump.

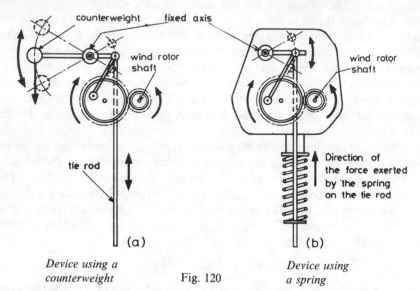

Device using a
counterweight   Fig. 120   Device using
a spring

5. DETERMINATION OF THE OPERATING CONDITIONS

The instantaneous pumping torque is periodic.   In calculating the mean torque that needs to be furnished by the windmill from the expression which relates it to the power ($P = C_r\overline{\omega}$), we obtain :

$$C_r = \frac{P}{\omega} = \frac{\overline{\omega}QH}{2\pi N}$$

and by substituting for Q, its value Q = kNq :

$$C_r = \frac{\overline{\omega}}{2\pi} kqH$$

q being the quantity of water supplied per stroke of the pump.

The mean torque is constant.   Its value is independent of the rotational speed and of the regulating mechanism.

The maximum value of the torque, on the other hand, depends on the crankshaft-connecting rod mechanism.

a) DETERMINATION OF THE MINIMUM WIND SPEED NEEDED FOR STARTING

It is the maximum torque which determines the starting wind speed.

Let us show on a graph (figure 121), the driving torque as a function of rotational speed for different wind speeds V_1, V_2, V_3. Their characteristics are represented by parabolas with their concave sides pointing downwards.

The horizontal line which represents the value of the starting torque C_d cuts the ordinate axis at a certain point D.

A characteristic " torque-speed " curve passes through this point which corresponds to a given wind speed V_0. This value is the maximum wind speed for which the windmill is at rest, under normal conditions. When the windspeed becomes higher than this value, the windmill starts. If no " torque-speed " characteristic passes through point D, the value of V_0 can be obtained by considering the intersection of the OC axis with the torque-speed curve (V_1), and by applying the following relation:
$V_0 = V_1 \sqrt{\dfrac{OD}{OB}}$ obtained from the similitude laws.

b) DETERMINATION OF THE SPEED OF ROTATION WHILE PUMPING

In order to determine the rotational speed of the windmill, one may use the torque-speed diagram or the power-speed diagram.

Torque-speed diagram

Let us draw on this diagram, the horizontal line whose ordinate is equal to C_r/η_p, η_p being the efficiency of the pump. The abscissas of the points of intersection of this horizontal line with torque-speed curves corresponding to different wind speeds give the corresponding rotational speeds of the windmill.

Power-speed diagram

We may also use the curves which represent the power developed by the windmill as a function of its rotational speed for different wind speeds (Figure 122). The abscissas of the points of intersections of these curves with the line of power demand $P'_{(N)}$:

$$P' = 2\pi N C_r/\eta_p = \overline{\omega} k N q H/\eta_p$$

give the rotational speeds of the windmill for the wind speeds considered. We may note that the starting wind speed can also be found by using the power-speed diagram. The magnitude of this wind speed is equal to the

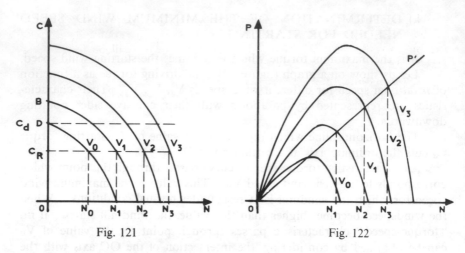

Fig. 121   Fig. 122

velocity corresponding to the wind speed curve tangent at 0 to the straight
line passing through 0 and whose equation is:

$$P = 2\pi\, N\, C_d$$

C_d being the starting torque calculated above.

In practice, it is preferable to use the torque-speed curves to determine
the starting wind speed in order to obtain a more accurate result.   The
power-speed diagram is, however, useful to determine the gear ratio n_1/n_2.

6. DETERMINATION OF THE OPTIMUM GEAR RATIO

Even the speed of low running windmills is usually faster than the
operating speed of the piston pumps which they drive.   Therefore, a gear
reducer is normally installed between the windmill shaft and the pump.

We have assumed in the preceding sections that the reduction ratio is
known.   In fact, it has to be fixed in such a manner that the best performance
will be obtained.

The gear reduction ratio has to be chosen in such a manner that the
operating points should fall in the neighbourhood of the maxima of the
power-speed curves.   In this region, the output of the windmill is at its
maximum.

In order to find the gear ratio, we may proceed in the following fashion:

Let us assume first that the pump is connected directly to the windmill.
We draw the straight line:

$$P_H = \overline{\omega} N' q H$$

which represents the power requirement of the pump as a function of its
rotational speed N', q being the amount of water pumped during one operating cycle of the piston.

In general, this straight line cuts the power-speed caracteristic curve of the windmill on its rising branch. Figure 123 shows that if we connect the pump directly to the windmill, the pump speed will be low. As a result, the water quantity supplied will be very small.

Let us assume that we have fixed the gear ratio k at a value other than one. The relation between the rotational speed N′ of the pump and that of the windmill N can be written as :

$$N' = \frac{n_1}{n_2} N = kN$$

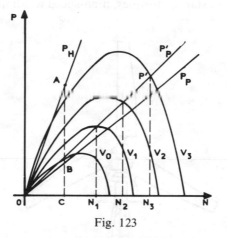

Fig. 123

Under these conditions, the ideal hydraulic power P put out by the pump for raising the water, and the power P′ which has to be supplied to the pump, can be expressed by the following relations :

$$P = \overline{\omega}N'qH = \overline{\omega}kNqH$$

$$P' = \overline{\omega}kNq\,\frac{H}{\eta_p}$$

Let us draw the two corresponding lines on the power-speed diagram, and choose the ratio k in such a manner, that the straight line, which represents the variation of P′ as a function of N, cuts the power-speed curves of the windmill close to their summits (Fig. 123).

The rotational speeds of the windmill are equal, on the average (the pumping torque not being constant), to the abscissas of the points of intersection of the straight line P′ with the windmill's characteristic curves.

In order to obtain the gear reduction ratio, it is sufficient to consider a point A on the straight line which represents P_H. The perpendicular line dropped from A onto the axis of abscissas cuts the straight line (P) in point B. The ratio $k = n_1/n_2$ is equal to the quotient of BC/AC.

Knowing the rotational speeds N_1, N_2, N_3, N_4 of the windmill at wind speeds V_1, V_2, V_3, V_4, lets the flow rate to be calculated as a function of the wind speed and a corresponding output curve to be drawn.

It is not difficult then, to calculate the amount of water being pumped per day, month or year.

C. Pumping of Water by High-Speed Windmills

1. INTRODUCTION

Due to their low starting torques, high-speed windmills are not suitable

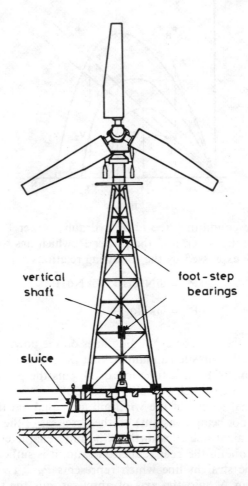

vertical shaft

foot-step bearings

sluice

Fig. 124 – *Fast wind turbine driging a propeller* pump for irrigation.

for driving piston pumps directly. Such an arrangement would not function correctly without inserting an electric transmission between the windmill and the pump, or a centrifugal clutch to allow the windmill to come up to speed before being connected to the load. This would result in mechanical complications and in an increase in weight for the installation.

It is better to associate a high-speed windmill with a centrifugal pump or with a helical pump having a fixed or variable pitch. The starting torque required by these pumps is quite low. Their rotational speed, on the other hand, is relatively high. As a result, they are well suited for coupling to high-speed windmills.

2. POWER-SPEED CHARACTERISTICS OF THE HIGH-SPEED WINDMILL

When the operation of high-speed windmills driving rotary pumps is considered, it is advantageous to use their power-speed diagrams.

We shall distinguish between two cases :

a) THE WIND SPEED IS HIGHER THAN THE RATED WIND SPEED OF THE MACHINE

The rotational speed of the windmill is being maintained at a constant level by means of an automatic device. Thus the windmill behaves as a constant speed electric motor.

b) THE WIND SPEED IS LOWER THAN THE RATED WIND SPEED OF THE MACHINE

The pitch of the blades is generally fixed. The rotational speed and the maximum power that can be developed increase with the wind velocity.

Figure 125 shows the power output of the windmill as a function of the speed of rotation for various values of the wind speed. The left hand side of the diagram corresponds to unstable operation. The summits of the curves correspond to maximum output. From the curve (V_1) representing the characteristics at wind speed V_1, another curve for wind speed V_2 may be deduced, by multiplying the abscissas of the different points of the curve (V_1) by the ratio V_2/V_1, and their ordinates by $(V_2/V_1)^3$. The point at the summit of the (V_1) curve becomes the summit of the (V_2) curve after the transformation.

Let N_1 and P_1 be the rotational speed and the power at the summit of the (V_1) curve. The corresponding values at the summit of the (V_2) curve are respectively :

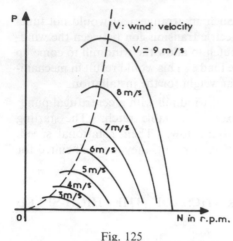

$$N_2 = N_1 \frac{V_2}{V_1} \quad P_2 = P_1\left(\frac{V_2}{V_1}\right)^3$$

By eliminating V_2 in these relations, we obtain :

$$P_2 = P_1\left(\frac{N_2}{N_1}\right)^3 = kN_2^3$$

The summits S of the curves (V) lie then on a cubic parabola which has the equation :

$$P = kN^3$$

Fig. 125

3. CHARACTERISTICS OF THE PUMP AND THE PUMPING STATION

A pumping test enables us to determine the drawdown y of the water level in the well as a function of the discharge.   When the diameter of the pipes is chosen, it is possible to calculate the manometric head Y for different discharges Q.

$$Y = h + y + \Delta H$$

h being the geometric height of the water lift and ΔH the losses of energy.

Fig. 127 represents the curve of variation of the quantity Y as a function of the water discharge.

On the same graph, let us plot the manometric characteristic of the pump H(Q) for a given rotational speed.   If the rotational speed of the pump is equal to the given value, the coordinates of the point of intersection define the water discharge and the manometric head corresponding to the operating conditions.

In fact, if the pump is driven by a windmill, the rotational speed varies except when the regulating system comes into action to keep it constant (for example, in case of a wind speed higher than the nominal value V_N). So we have to deal not only with a single characteristic but with several.

It is possible to obtain the other characteristics from the first by applying the classical relationships of similitude relative to hydraulic machines.

$$\frac{Q_2}{Q_1} = \frac{N_2}{N_1} \quad\quad \frac{H_2}{H_1} = \left(\frac{N_2}{N_1}\right)^2$$

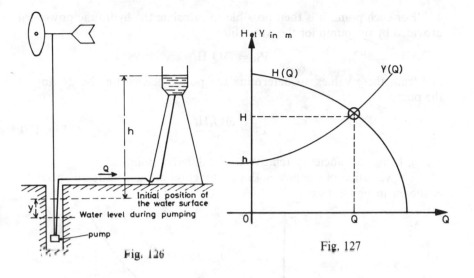

Fig. 126

Fig. 127

Let us represent these characteristics for various rotational speeds: N'_0, N'_1, N'_2, N'_3, N'_4...

The characteristic $Y(Q)$ of water installation which can be plotted on the same graph, cuts the characteristics $H(Q)$ in several points: M_0, M_1, M_2, M_3, M_4, etc...

The coordinates of these points give the discharges which would be extracted and the corresponding manometric heads for the various rotational speeds: N'_0, N'_1, N'_2, N'_3, N'_4.

N'_0 being the rotational speed for which the pump begins to extract water from the borehole.

Thus we can draw the variation curve of the discharge Q as a function of the rotational speed N' of the pump (fig. 129).

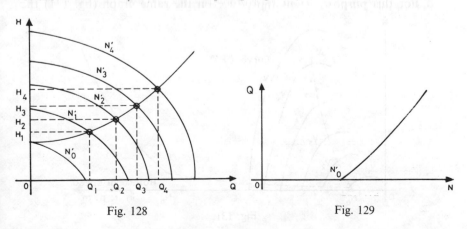

Fig. 128

Fig. 129

For each point, it is then possible to calculate the hydraulic power P_H provided by the pump for the water lift:

$$P_H = \overline{\omega} Q_i H_i$$

and from these values, to determine the power needed for the driving of the pump.

$$P_i = \frac{\overline{\omega} Q_i H_i}{\eta_i}$$ Fig. 130.

η_i being the efficiency relative to the different points M_i.

The variation of the power P_i as a function of the rotational speed N' is shown in figure 130.

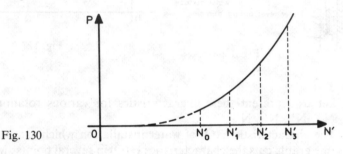

Fig. 130

4. DETERMINATION OF THE GEAR BOX RATIO AND THE OPERATING CONDITIONS

The pump shaft is connected to the windmill shaft through a gear box. As a consequence, the rotational speed of the pump is proportional to that of the wind turbine. The problem is to choose the best gearing up ratio for optimal conditions.

For this purpose, let us reproduce on the same graph (fig. 131) the

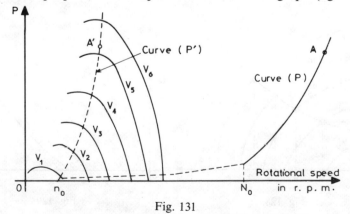

Fig. 131

power rotational speed characteristics of the wind rotor and that of the pump (curve $P(N_1')$ as previously defined).

If the wind turbine shaft drove the pump shaft directly, there would be no water extracted, the pump beginning to deliver water only when its rotational speed reaches N_0'.

Thus, it is necessary to insert a gear box between the pump and the wind turbine.

Assume the gearing up ratio k is known.

To determine on the graph the points corresponding to the operating conditions for the various wind speeds V_1, V_2, V_3 etc..., we can multiply by k the abscissas of the characteristics of the wind rotor and by energetic efficiency η_e of the gears, its ordinates. The points of intersection of the obtained characteristics with the $P(N)$ curve give the operating conditions.

In fact, it is simpler to keep the wind turbine characteristics unchanged and to reduce the abscissas of the power-rotational speed of the pump by multiplying them by $1/k$ and to multiply its ordinates by the ratio $1/\eta_e$

The coordinates of the points of intersection of the new characteristics P′ so determined with the wind rotor characteristics give the power supplied and the speed of rotation of the wind turbine for the various wind velocities V_1, V_2, V_3 etc...

It is desirable that the points of intersection be located near the apex of the power-rotational speed characteristics of the wind turbine. If so, it means that the wind power is fully used for water pumping and the wind rotor is working at its maximum efficiency.

For determining the ratio k, we can choose a point A′ of the P′ curve on the cubic parabola which joins the apex of the characteristics of the wind turbine. If A′ is the corresponding point of the curve P, the ratio abscissa of point A/abscissa of point A′ is equal to the value of the gearing up ratio which has to be adopted. We can then determine each point of the P′ curve by applying the above rule to the P curve.

The coordinates of the points of intersection of the P′ curve with the power-rotational speed of the windmill for the various wind velocities, give us the power provided and speed of rotation of the wind turbine and, therefore, the rotational speed of the pump.

The water discharge depends on the speed of rotation of the pump which is proportional to that of the wind rotor. As this last rotational speed is itself a function of the wind velocity in the previous operating conditions, it is possible to draw the variation curve of the water delivery as a function of the wind speed.

Only one value of the water discharge corresponds to each wind velocity.

The intersection of the Q(V) curve with the abscissas axis OV shows the wind velocity for which the pump begins to supply water.

For the wind velocities higher than the nominal wind speed V_N, the water discharge keeps the value corresponding to V_N.

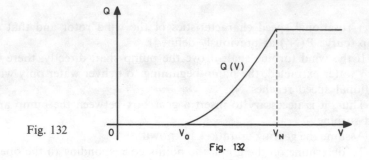

Fig. 132

5. DETERMINING THE WATER VOLUME EXTRACTED

The knowledge of wind velocities, obtained from wind anemometer records, enables the determination of the water volume extracted daily, monthly or yearly.   This calculation is made easier by use of the monthly and yearly speed-duration curves.

We plot (fig. 133) in the plane, divided into 4 parts :

— in the first quadrant : the variation curve of the water discharge as a function of the wind velocity,

— in the fourth quadrant : the wind speed-duration curve during the considered period (month or year) with the nominal and maximal wind velocity V_N and V_M whose definitions were given in chapter 1.

Then we construct in the second quadrant, one point after the other, the water discharge-duration curve i.e. the curve giving the water discharge in m^3/day as a function of the number of days during which a higher value of discharge can be obtained.

The water volume extracted is proportional to the hatched area limited by this curve in the second quadrant.

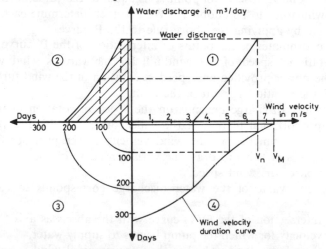

Fig. 133 – *Determining the volume of water delivered.*

Note that in the event of an insufficient water volume, it would be necessary to come back to the measurements determining the windmill and to the choice of the pump.

6. DETERMINING THE WATER-STORAGE

We can manage in two ways :
— by considering the duration of the periods during which the wind is unproductive,
— by drawing the cumulative volume of the extracted water and the needed water as a function of time.

a) ACCOUNTING FOR THE DURATION OF THE PERIODS OF UNPRODUCTIVE WIND

Below a certain velocity, the wind power is not sufficient for pumping water from a well or a borehole.

From the curve giving the number of the periods as a function of their duration during which the wind has been unproductive, it is possible to calculate, approximately, the volume of the water tank.

First hypothesis : In that, the water tank provides water to the village continuously. The curve (fig. 134) shows that its volume has to be equal to the water consumption for eight days.

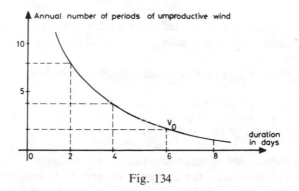

Fig. 134

Its volume will then be calculated by the expression :

$$V = 8nv$$

n being the number of inhabitants and v the daily per capita water consumption in m^3.

Second hypothesis : If the calculated volume seems to be excessive, we shall determine the size of the watertank needed for a shorter period, for example : for 4 days, but it should be noted that the water supply will be insufficient in the village, at least :

      — once a year during 4 days,
      — twice a year during 2 days,
      — three times a year during 1 day.

b) USING THE CUMULATIVE VOLUME CURVES OF THE WATER EXTRACTED AND THE WATER NEEDED

Remember that the cumulative volume curve is the variation curve of the quantity :

$$V = \int_0^t Q dt$$

as a function of time

$$V = V_e \quad \text{and} \quad Q = Q_e \quad \text{for the extracted water}$$
$$V = V_d \quad \text{and} \quad Q = Q_d \quad \text{for the needed water.}$$

The curve $V_e(t)$ represents the sum of the daily volumes capable of being extracted from the instant t_0 to the present t, assuming that the pumping begins at the instant t_0, and the curve $V_d(t)$, the quantities of water required to satisfy the needs of the population during the same period.

The curve $V_e(t)$ is constructed from anemometer records. For each wind speed, there is a corresponding water discharge which can be known from the curve $Q(V)$.

The quantity $V_d(t)$ depends on the number of inhabitants and on the use of water.

If we considered that the water need is constant, then :

$$V_d(t) = nv(t - t_0)$$

n being the number of inhabitants and v, the daily capita consumption.

Figure 135 represents the curves $V_e(t)$ and $V_d(t)$ and figure 136, the difference between the ordinates of these two curves as a function of time. If we want to have water without a break, the volume of water tank to be adopted is equal to the highest value V_1, V_3, V_3 as shown in figure 136. V_1, V_2, V_3 are the differences of the ordinates in figure 136 between a maximum and the lower minimum which follows.

During the strongest wind velocities, in order to avoid overspeeding of the wind rotor, we have to be sure that the water level in the well will permanently remain higher than the inlet of the aspiration water duct.

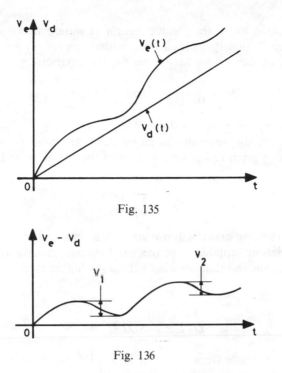

Fig. 135

Fig. 136

7. APPROXIMATIVE PRACTICAL VALUE OF THE DISCHARGE ABLE TO BE EXTRACTED BY SLOW AND FAST WIND TURBINES

Assuming that the wind power plant is correctly determined.

The maximal powers which can be provided by slow and fast wind turbines are given by the relations :

$$P_L = 0.15 \ D^2 V^3 \qquad \text{and} \qquad P_R = 0.2 \ D^2 V^3$$

For reasons we have discussed at the beginning of this chapter, the slow wind turbines are generally coupled to piston pumps and the fast wind rotors to centrifugal or helical pumps.

Piston pumps generally have a better efficiency than centrifugal pumps. We can assume an efficiency of 65 % for the piston pump and the mechanical drive and 50 % for the centrifugal pump and its connection to the wind rotor. The discharges capable of being extracted by using slow and fast wind turbines are then, such as :

$$\overline{\omega} Q_L H = 0.65 \ P_L \qquad \text{and} \qquad \overline{\omega} Q_R H = 0.50 \ P_R$$

Q_L and Q_R being the discharges obtained by using respectively slow and fast wind rotors.

Substituting for $\bar{\omega}$, the specific weight of water, the value 9 800 N/m^3 and replacing P_L and P_R by their expressions as a function of diameter and wind speed, we can obtain for Q_L an Q_R the relations:

$$Q_L = \frac{D^2 V^3}{10^5\, H} \quad \text{and} \quad Q_R = \frac{D^2 V^3}{10^5\, H}$$

Q_L and Q_R being expressed in m^3/s.

In practice, the water discharge whatever the wind rotor type may be (slow or fast) is given in l/s as a function of the diameter by the single relationship:

$$Q = \frac{1}{100} \frac{D^2 V^3}{H}$$

D and H being expressed in meters, V in m/s and Q in l/s.

This relation supposes that the wind turbines rotate in the best operating conditions and that the wind velocity is higher than the starting wind speed.

OTHER SOLUTIONS

To extract water from a well, it is possible to use other systems than piston or centrifugal pumps. In particular, we can use pumps of the screw type, diaphragm pumps or produce compressed air or electricity with the windmill for pumping water.

Diaphragm pumps are used for small wind plants when the water lift required is not very high. This condition is not very often met and these pumps are not very robust. In the following pages, we shall study water pumping by means of screw pumps, compressed air and by generating electricity.

1. UTILIZATION OF A SCREW PUMP

This type of pump is not very widespread. However, nowadays, it is more often used because of its robustness. Similar equipment is used for pumping viscous liquids.

The screw pump consists of a screw which rotates in a rubber stator or of two endless screws which gear one against the other.

The screw pump is a volumetric pump. It opposes a constant torque to the windmill shaft rotation. Its discharge only depends on the rotational speed and not on the water lift. The efficiency of the screw pump may reach 75 or 80 % for a water lift of about 30 m. The rotational speeds vary between 100 r.p.m. and 1 000 r.p.m. depending on the type. Screw pumps can be coupled with slow or fast wind turbines.

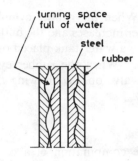

turning space
full of water

steel

rubber

Fig. 137

Experimentation shows that the wind turbines which have a tip-speed ratio less than 3 are more effective.   They start in low speed winds at about 4 m/s because of the higher starting torque of these machines.

In practice, to facilitate the starting, a by-pass device can be used which empties the water lift duct before starting, or a centrifugal clutch which drives the pump as soon as the rotational speed is high enough. To equip deep boreholes, it is better to use pumps with small diameters and high rotational speeds.

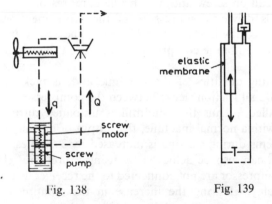

q

Q

screw
motor

screw
pump

elastic
membrane

Fig. 138   Fig. 139

2. PUMPING BY MEANS OF HYDRAULIC DRIVE

When the well is too deep, in order to avoid a heavy mechanical drive, we can use a hydraulic drive as shown in fig. 138.   The pump, which is fastened on a high-speed windmill, feeds fluid into the hydraulic motor which is connected to the water lifting pump.

With multicylinder axial piston pumps and motors, the efficiency can reach 50 % or 60 % and with screw pumps and motor 40 % or 50 %.

We can also use a volumetric pump such as a Vergnet pump.   An alternating hydraulic motor driven generally by a slow wind turbine feeds fluid periodically into a dilatable rubber cylinder placed in a steel cylinder fitted with inlet and outlet valves. When the rubber cylinder dilates, the water pressure inside the steel cylinder increases and the outlet valve opens, allowing the water to be directed into a water tank placed outside the well, on the soil surface (see figure 139). When the rubber cylinder reduces the outlet valve shuts and the inlet valve opens, allowing the water to be introduced into the steel cylinder.

3. USE OF A HYDROEJECTOR

This device is the result of the combination of a centrifugal pump located outside the well and a Venturi pipe placed inside the borehole, below the water pumping level. A part of the water is forced back by the pump through the aspiration duct. The water pipes must be filled before starting to prime the system.

4. PNEUMATIC DRIVE

Water pumping is also possible by using compressed air. To produce this, multi-piston compressors with a crosswise arrangement may be used. We can also call on screw and membrane compressors.

Windmills driving compressors of the afore mentioned types were constructed especially in the Soviet Union. The Soviet experiments show that the multi-membrane compressor fastened on the windmill shaft is the best solution.

The starting qualities of the machine and its pick-up are improved when a centrifugal friction sleeve between the compressor and the windmill shaft is installed. Thus the windmill is unloaded during starting. By comparison with a normal machine, the output of machines equipped with relieving systems during starting is increased by 3-5 times.

This aim can also be achieved by using a pressure relay : The cylinders of the compressor are not connected to the receiver directly but through a pressure relay. During the increase in the compressor productivity, the air is directed into the receiver. When the rotational speed of the wind turbine decreases, the relay disconnects the compressor cylinders from the receiver and connects them to the atmosphere.

For the pumping itself, many possibilities may be used :
— lift of the water by means of an emulsion,
— centrifugal pump driven by a compressed air motor fed by the compressed air produced by the wind plant,
— displacement of the water by the compressed air.

a) WATER PUMPING BY EMULSION

Let h be the height of the water lift.
To be in the best conditions for pumping water, the pipe containing

the emulsion of air and water must have a length of 2 h and be sunk into the water by a depth equal to h.

The efficiency is less than 30 %.

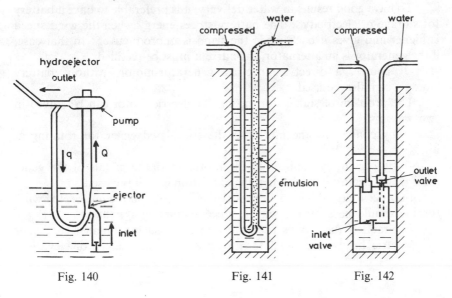

Fig. 140   Fig. 141   Fig. 142

b) DISPLACEMENT WATER PUMPS

These mainly consist of a cylinder sunk into the water which has to be pumped. At the base of the cylinder, an inlet valve is found. In the absence of compressed air, the inlet valves opens and water penetrates into the pump. The air admitted through an automatic device forces the water out of the cylinder into the lift pipe through an outlet valve.

In some models, the compressed air, after being released, passes into the water lift pipe. In other ones, the compressed air, after release, goes to the atmosphere or returns towards the inlet valve of the compressor.

The deficiency in using a diaphragm and piston compressor is that the power required for constant pressure is proportional to the rotational speed. They oppose a constant mean torque to the rotation of the rotor shaft. Therefore, it is impossible to utilize fully the power of the wind rotor.

5. ELECTRICAL WATER PUMPING

The wind turbine, generally a fast running one, drives an electric generator which feeds an electric motor connected to a pump. This solution is valuable when aquifers are deep under the soil surface, and when the mean wind speed is higher than 5 or 6 m/s.

The erection of the wind-driven generator away from the water source, at an elevated place, where the wind rate is higher, can increase the power output by 30 % or more.

To have good results in water delivery, it is preferable to have a battery for storage of electricity. The battery stores energy when the wind speed is higher and gives it back when the wind is unproductive. In that case, if the generator is an alternator, the current must be rectified.

Of course, a direct connecting generator-motor, without battery storage, can also be used.

The problem of stable starting of the electric motor can be solved in two ways :

— by connecting the motor to the stimulated generator rotating at idle,

— by previously connecting the motor to the terminals of the generator which begins to work when the excitation circuit is closed.

In order to avoid destruction of the generator and the motor, the wind turbine should be provided with a speed regulating system.

In practice, the pumps used are chosen in such a manner that their manometric heads, are equal to 1.5 or 2 times the height of elevation.

CHAPTER VII

THE GENERATION OF ELECTRICITY
BY WIND POWER

In the windy parts of the world, the production of electricity by wind turbines may be very economical. It is particularly useful for isolated and remote villages.

In the present chapter, we deal with electrical installations and different problems which occur in electricity generation designs such as the generator-turbine coupling, the determination of gear-box ratio, the electrical connection to the grid etc... Then we describe several installations having a capacity from 500 W to 5 000 kW.

A. Installation Designs

1. THE WIND TURBINES

The wind turbines used for generating electricity are generally high-speed machines having two or three blades for the following reasons:

— At equal diameter, the high-speed wind machines are lighter, therefore cheaper than the low-speed ones.

— They rotate at a higher speed. Thus the necessary gear-up ratio is lower. Consequently, the step-up gearing is lighter.

— The necessary torque to start an electrical generator is very low. Though the starting torque of the fast wind rotor is itself very low, it is sufficient to drive the generator into rotation. Therefore the high-speed wind turbines are well-suited to this use. Nevertheless, some low-speed machines have also been tested.

The machines have fixed or variable-pitch blades. In some designs, starting is made easier by a special regulator which increases the pitch when the rotor is stopped.

In others (high-speed wind machines with fixed blades), the generator acts as a motor during starting and turns into a generator when the normal speed of rotation is reached.

There are also wind rotors which have no special systems. For these wind rotors, starting is more difficult, especially when the value of the tip-speed ratio λ_0 of the machine is very high. Such machines have to be equipped with twisted blades.

Generally, the wind rotor drives the electrical generator through a step-up gearing. The improvements made in the construction of gearing and the high price of electrical generators rotating at low speed tend towards the elimination of the direct drive of the generator by the rotor, except sometimes for small units.

As for pumping, the wind rotor may be located upwind or downwind of the tower. In the first case, the effect of wake is avoided. In the second one, as the torque required for orientation is lower, the power of the yawing motor is reduced.

2. ELECTRICAL GENERATORS

Three types of generators are commonly used: the direct-current generator (shunt type), the alternator (or synchronous generator), the induction generator (or asynchronous generator).

In small power installations, direct-current generators, which were much used formerly, are now often replaced by synchronous or asynchronous generators. These can provide alternating current which can be easily transformed into direct current by means of rectifiers which are very inexpensive.

In medium and high power installations alternators and induction generators are the most wide-spread.

The advantage of the alternator relative to the direct-current generator lies in its efficiency, which is higher, and in the fact that the alternating-current generator can provide electricity at a lower speed of rotation than a direct-current generator. It can also supply electricity at higher speeds. The ratio between the maximum speed of rotation and the minimum speed necessary for the production of electricity is higher for an alternating-current generator. Thus a wind turbine driving an alternator will be able to use a wider range of wind speeds.

With respect to the induction generator, the alternator has the advantage of supplying its own magnetizing current but it is more expensive. The connecting control system is more sophisticated. It consists of a tachometer, a voltmeter, a phasemeter, an automatic device to make the

15. Aéroturbine wind generator (France).

16. Savonius rotor (I.U.T., Dakar, Sénégal).

17. MOD.1 wind turbine (USA).

18. MOD.2 wind turbine (USA).

19. Windmatic windmill (Denmark).

20. Kuriant windmill (Denmark).

21. Nibe wind turbine, model A (Denmark).

22. Tvind aerogenerator (Denmark).

23. Dansk Vinkraft gyromill (Denmark).

24. Poulsen windmill (Denmark).

25. Gotland wind generator (Sweden).

26. Maglarp wind generator (Sweden).

27. Aeroman wind generator (W Germany).

28. Brümmer wind generator (W Germany).

29. Orkney wind turbine (Great Britain).

30. Growian I (W Germany).

connection to the grid and a reverse power relay ensuring disconnection if the wind drops or if the voltage of the grid falls to zero. Connecting the synchronous generator to the network requires precise adjustment and raises a serious problem as this operation has to be performed frequently. The machine has to be switched in exactly at the synchronous speed, the voltage of the alternator being in phase and equal to the voltage of the grid. It is possible to overcome the difficulties by the use of large dampers. These dampers, special coils in the shape of a squirrel cage, allow the synchronous generator to start like an asynchronous motor, and suppress or reduce the oscillations which may occur when the connection to the grid is made. Another solution which has been suggested consists of interposing, between the wind turbine and the alternator, a free-running coupling leaving the generator connected permanently to the grid. But this solution is not economical.

The asynchronous generator seems to be the cheapest and most reliable solution. It has several advantages :
— Its construction is inexpensive,
— No rotating contacts, so starting is easy,
— Ease of connection to the network,
— Absence of oscillations when it is coupled to the grid.

The induction generator may be connected to the grid with a speed different by several per cent from the synchronous speed without inconvenience, the resulting overload being of very short duration. The connecting control system consists only of a tachometer contact controlling the switching-in to the network and a reverse current relay ensuring disconnection if the wind drops.

A small disadvantage of the asynchronous generator is that it takes its magnetizing current from the grid and absorbs reactive power. But this small inconvenience may be overcome by connecting well-fitted capacitors to its terminals. The use of such elements reduces or suppresses the supplying by the grid of reactive power and therefore improves the power factor.

In Denmark, the induction asynchronous generator is favoured by manufacturers because it can be used as a motor for starting the machine. If the wind velocity is higher than the cut-in speed, it becomes a generator of electric power when it reaches its normal speed of rotation. If the wind speed is lower than V_m, an automatic device disconnects it from the grid. This system is very worthwhile for high speed fixed-blade machines whose starting torque is low.

In the latest high power installations and designs, the use of induction asynchronous generators is the solution generally chosen.

Determination of the operating conditions

In the following paragraphs, the operating conditions are examined.

Two possibilities have to be considered :

— The generator driven by the turbine feeds into a direct-current network or into an alternating-current network where it constitutes the only power unit.   Its rotational speed is variable.

— The generator feeds into an alternating-current network at fixed frequency.   Its rotational speed is constant or nearly constant.

3. THE GENERATOR SUPPLIES POWER TO A D.C. NETWORK OR TO AN A.C. NETWORK WHERE IT CONSTITUTES THE ONLY AVAILABLE POWER UNIT

In this case, the problem of generating electricity is not very different from that of pumping water.   Instead of driving a pump which provides a certain discharge under a given manometric lift, the wind turbine drives an alternator which supplies current under a determined voltage.

To solve the problem, the windmill, generator and network characteristics must be plotted.

a) WINDMILL CHARACTERISTICS

In a previous chapter, the characteristics have been presented, so we shall not repeat them.

b) OUTPUT CHARACTERISTICS OF THE GENERATOR

We mean the characteristics $V(I)$ voltage/intensity for different speeds of rotation N_1, N_2, N_3, N_4 (fig. 143).

c) LOAD CHARACTERISTIC

The load characteristic $V = f(I)$ depends on the network structure.

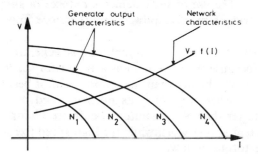

Fig. 143 — *Generator and network characteristics.*

This curve cuts the output characteristics of the generator in several points whose ordinates enable us to determine the current intensity, the voltage and the power delivered for the different values N of the rotational speed.   It is possible to draw the variation curve of the power P(N) supplied by the generator versus the speed of rotation.

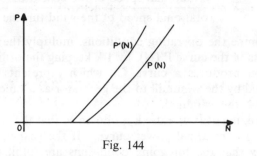

Fig. 144

The mechanical power P'(N) provided by the wind turbine to drive the generator may be obtained from the former curve by adding to its ordinates, the corresponding losses of the generator (losses by Joule's effect, mechanical and magnetic losses) and also the mechanical losses in the step-up gearing.

d) CHOICE OF THE GEAR-UP RATIO.   DETERMINATION OF THE OPERATING CONDITIONS

The gear-up ratio may be determined exactly as in the case of a centrifugal pump driven by a wind machine.

On the same graph (fig. 145) the power-rotational speed curve and the previous characteristic P'(N) is drawn.

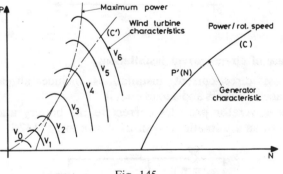

Fig. 145

If the wind turbine and the generator were directly coupled, the former would rotate like a rotational anemometer and the power delivered by the generator would be zero because the rotational speed of the latter is too low.   To produce power, the rotational speed of the shaft driving the generator has to be increased through a step-up gearing.

Let the gear-up ratio be k.   Take it as a known factor.

$$k = \frac{\text{rotational speed of the generator}}{\text{rotational speed of the wind turbine}}$$

To determine the operating conditions, multiply the abscissas of the different points of the curve $P'(N)$, by $1/k$ keeping the ordinates unaltered. This operation produces a curve C', which represents the mechanical power provided by the windmill to the generator as a function of the rotational speed of the windmill.

In practice, the gear-up ratio k is chosen so that the curve C' is as near as possible to the maximal power curve.   If the gear-up ratio is chosen in such a way that the foregoing conditions are fulfilled, the efficiency of the wind energy conversion system will be maximal.

The intersections of the curve C' with the power rotational speed characteristics of the wind turbine enable us to get the power P' supplied by the wind turbine, and its rotational speed for the various wind speeds. Thus, it is possible to draw a curve $P'(V)$ and to deduce from it a curve $P(V)$ giving the electrical power delivered by the aerogenerator for the different wind velocities.

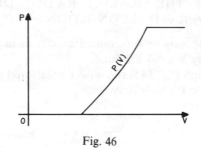

Fig. 46

Particular case of direct-current installations

Generally, direct-current installations include an aerogenerator, a battery of accumulators and a load circuit.

The aerogenerator provides current to the battery and to the load circuit through an automatic cut-out as shown in fig. 147.

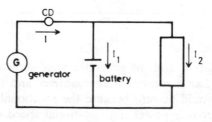

Fig. 147

a) BATTERY CHARACTERISTIC

When the wind speed is sufficient, the generator voltage exceeds that of the battery. The cut-out contacts close and the battery charges. The battery voltage varies according to the following expression :

$$V = E + rI_1$$

If the generator voltage is lower than the battery voltage, the cut-out contacts remain open. The battery discharges and the voltage to its terminals varies with the current I_1 according to the law :

$$V = E - rI_1$$

This expression may be written as :

$$V = E + rI_1$$

provided one considers that I is positive if the battery charges and negative in the other case.

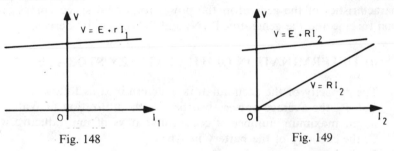

Fig. 148   Fig. 149

b) LOAD CHARACTERISTIC

If the load consists only of a resistance, the load characteristic will be a straight line :

$$V = RI_2$$

If the load consists of a direct current motor, the above expression is replaced by the following one :

$$V = E + RI_2$$

c) NETWORK CHARACTERISTIC

The network to which the generator supplies power includes the battery and the load. When the cut-out contacts are closed, the generator voltage is equal to the battery and load voltages. Moreover, between the different currents, the following relationship may be expressed as :

$$I = I_1 + I_2$$

As a result, the network characteristic $V = f(I)$ is obtained by adding, at equal ordinates, the abscissas of the different points of the load and battery characteristics (fig. 150).

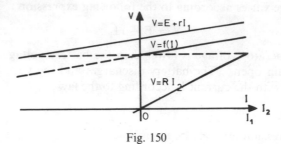

Fig. 150

Thus we are brought back to the general case studied in an earlier paragraph.   From the intersections of the curve $V = f(I)$ with the output characteristics of the generator, the power/rotational speed curves of the wind turbine and the generator, $P'(N)$ and $P(N)$, may be derived.

d) DETERMINATION OF THE BATTERY STORAGE

The capacity of the accumulators is determined as follows :
Call P_m the average power absorbed by the utilization in Watts,
n, the maximum number of consecutive days of unproductive wind,
C, the capacity of the battery in Ah,
E, the battery storage in volts.
To avoid an interruption in the electric current supply and a complete discharge of the battery, we must have :

$$0.8 \, CE > 24 \, n \, P_m \quad \text{thus:}$$

$$C > \frac{24 \, n \, P_m}{0.8 \, E} \quad \text{or} \quad C > \frac{30 \, n \, P_m}{E}$$

Example : $E = 30 \, V$ $P_m = 20 \, W$ $n = 10$

$$C > 30 \, \frac{10 \times 20}{30} = 200 \, \text{Ah.}$$

Besides the above condition, in order to avoid the battery being damaged, the maximum current I_M during the charge by the generator must not exceed $C/10$.   Thus the capacity of storage C will also have to meet the condition : $C \geqslant 10 \, I_M$.

4. THE GENERATOR FEEDS INTO AN ALTERNATING NETWORK INCLUDING OTHER POWER UNITS AT A CONSTANT FREQUENCY

a), THE GENERATOR IS AN ALTERNATOR

Its rotational speed remains constant. It is the same as that of the wind rotor if the gear-up ratio itself remains unaltered.

Fig. 151 illustrates the power/rotational speed characteristics of the wind turbine and of the alternator (full curves).

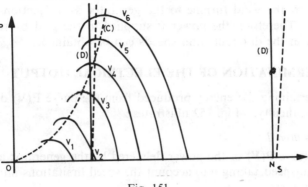

Fig. 151

The characteristic P(N) of the alternator is a vertical straight line (D) perpendicular to the horizontal axis at a point whose abcissa is N_S, N_S being the synchronous speed.

If k represents the gear-up ratio, the rotational speed of the wind turbine is N_S/k. Let (D') be the vertical line whose abscissa is N_S/k.

The ordinates of the intersections of (D') with the power rotational speed curves of the wind turbine enable us to get the values of the power supplied by the windmill to the generator for different wind speeds.

In practice, the gear-up ratio will be chosen in such a manner that the above points of intersection will be near the summits of the power characteristics of the wind turbine, especially for the most productive wind velocities.

If the mechanical, magnetic and electrical losses of the step-up gearing and the alternator are known, it is possible to calculate the power P provided by the alternator to the grid at the various wind speeds.

Note that the whole set is unproductive under a certain wind speed equal to V_2.

b) THE GENERATOR IS AN INDUCTION GENERATOR

Its rotational speed increases slightly with the load. There is a slip relative to the synchronous speed without however exceeding 5 to 6 %. In this case, the characteristic P(N) of the generator is a nearly vertical dotted curve (C), as shown in fig. 151.

The gear-up ratio and the efficiencies of the step-up gearing and generator being known, it is possible to obtain from the curve (C) the characteristic P′(N) (dotted curve (C′)) giving the power supplied by the wind turbine to the generator as a function of the rotational speed N of the wind rotor.

The intersection of this curve with the power/rotational speed characteristics of the wind turbine allows the determination of the power P′ provided by the wind turbine to the generator as a function of the wind speed V. Therefore, the power P supplied to the grid by the induction generator at the different wind speeds can be obtained.

5. DETERMINATION OF THE ELECTRICAL OUTPUT

Determining the energy produced from the curve P(V) does not present any difficulty.   Fig. 152 illustrates :

— *in area 1 :*

the curve P(V) of the power delivered by the generator as a function of the windspeed, taking into account the speed limitations imposed by the regulating system,

— *in area 4 :*

the speed duration curve,

— *in area 2 :*

the power-duration curve derived from the above curves.

The area situated between the axis in area 2 and the power curve is proportional to the annual energy output.

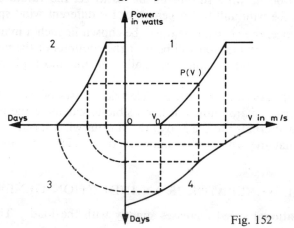

Fig. 152

6. REMARKS ABOUT REGULATING SYSTEMS

Whatever electrical solution is adopted, mechanical regulation of the power delivered is necessary. This regulation can be done either by adjusting the blades or by an aerodynamic brake.

Wind machines driving a direct-current generator or an alternator feeding into an autonomous grid, whose frequency is imposed by the generator itself, can be regulated by tachometers.

For aerogenerators with variable-pitched blades feeding into a grid of constant frequency, power regulation is much better. Often the mechanical speed regulator will also serve to limit power output. It will also provide speed limitation when the generator is disconnected from the grid.

For the fixed blade machines feeding into a constant frequency grid, speed regulation is not necessary because the grid imposes the speed of rotation on the wind rotor. In this case, power regulation is obtained almost simultaneously : when the wind speed increases, the tip-speed ratio decreases because of the constant value of the rotational speed. Thus the efficiency diminishes and the power is not as high as it would be if the tip-speed ratio kept a constant value. The power limitation is due to the fact that blade tip sections are working near their stalling point.

Thus a machine designed for a rated speed V_N and for a tip-speed ratio $\lambda_0 = 6$ will operate with a tip-speed ratio $\lambda_0 = 3$ if the wind speed becomes $2\ V_N$ because of the constant value of the rotational speed, but with a reduced efficiency.

However, to avoid racing if the generator is disconnected from the grid, a fixed-blade machine will have to be provided with a braking system : a mechanical brake on the shaft and an aerodynamic braking system at the tip of the blades.

For autonomous fixed-blade machines, the regulating effect can be obtained by using hypercompound generators which can provide power increasing proportionally to the cube of the speed of rotation, or asynchronous generators shunted by static capacitors supplying a load shunted by variable electrical resistors, electronically controlled.

7. ENERGY STORAGE

Wind power is a very irregular energy source, hence energy storage is necessary. Several possibilities exist, but none of them is perfect.

Thermal storage :

Thermal storage may take several forms : Heating of water, heating of gravel and stones in an insulated tank or melting of substances which give up latent heat when they return to their former state. The stored heat is then generally used for space heating.

Water pumping:

This system, which may be found in some hydraulic power designs has never been used for wind energy conversion. Water is pumped into an elevated tank or reservoir and then used to turn a turbine as energy is needed (efficiency: 60 % to 80 %).

Inertia storage:

Storing energy by means of a fast-running fly-wheel is not a new idea. As early as 1950, the Swiss had Oerlikon buses propelled by energy stored in fly-wheels (gyrobus).

Recently, an American university proposed to make fly-wheels of composite material (metal + polyester resin). However, the possibilities of energy storage remain limited because, beyond a certain rotational speed, the wheel may explode. With a fly-wheel running at 15,000 r.p.m. on magnetic bearings, it is theoretically possible to store 400 Wh/kg for 24 hours. The efficiency of the system (restored energy/consumed energy) is excellent (about 80 %).

Storage by compressed air:

Compressed air is forced into a manufactured tank or into an underground anticlinal chamber. Later, it may be used in two manners: either by direct expansion through a compressed air motor, or by injection into an internal-combustion turbine in which the oxygen it contains is burnt with fuel in a combustion chamber to supply mechanical energy. Respective efficiencies: 60 % and 80 %.

Hydrogen storage:

Hydrogen is generated by electrolysis of water by the direct current provided by the aerogenerator. This may then be highly compressed and stored in cylinders, or stored at low pressure in a gas holder. It can be used for heating or cooking or to run an engine.

Another possibility, after compression, is to inject hydrogen into a fuel cell which directly converts chemical energy into electrical energy when it is needed.

Efficiency: 60 to 70 %.

Accumulators:

Electrical accumulators are commonly used for storing energy.

The best batteries are lead-acid accumulators. They are well-suited for trickle charging. In quantity of electrical output, their efficiency is about 80 % to 90 %. In energy, it is about 70 % to 80 %. For large installations, batteries with thick plates are used. For small installations, accumulators such as those used in trucks are sufficient. The main causes of rapid deterioration are overcharging, overdischarging, and being left in a discharged state.

Nickel-cadmium batteries are not recommended because their effi-ciency is very low at small intensities and lower then those of lead-acid accumulators at any intensity, However, they are not damaged by overcharging nor by occasional overdischarging, have no self-discharge and are less liable to damage by frost than the lead-acid type.

8. LIGHTNING PROTECTION

Aerogenerators are often mounted on towers or pylons erected on hilltops. With their supports they constitute ready paths for transfering static electricity from the clouds towards the earth.

To prevent damage by lightning, the pylon, which sustains the gene-rator, has to be connected to the earth by good conducting wires.

For composite blades, special lightning protection is needed. The recommended provisions for large composite blades consist of a full chord metal tip cap having an 8 or 10 cm skirt extending inboard from the tip. A trailing edge earth strap must be firmly attached to the tip cap and to the steel hub adapter to carry the lightning current to the ground. Metal screening, conductive paint, or other conductors must be disposed along the blade to ground lightnings which strike neither the blade-tip cap nor the trailing edge. Similar materials have to be added especially in the areas which correspond to internal metal parts to preclude lightning stroke penetration through the composite spar.

The above recommendations are those proposed by H. G. Gewehr of the Kaman Aerospace Corporation who has studied lightning protection for the blades of the big American wind turbine MOD. 2.

B. Small Wind Power Plants

In this section, we shall examine the wind power plants whose diame-ters are inferior to 20 m. Some of the wind machines described here are experimental units. Others are mass-produced by industry.

1. FRENCH WIND TURBINES

Paris-Rhône wind turbines (fig. 153)

The wind rotor is composed of two main fixed blades made of wood and two auxiliary blades of variable pitch. The latter provide high torque for starting, but if the rotational speed becomes too great, the centrifugal force acting on them reverses their pitch angles, causing them to act as a brake. The machine is oriented by a tail vane.

The wind power plants constructed by the Paris-Rhône Company, include a 500 W mode 1. This wind turbine drives a direct-current gene-

Small French wind generators

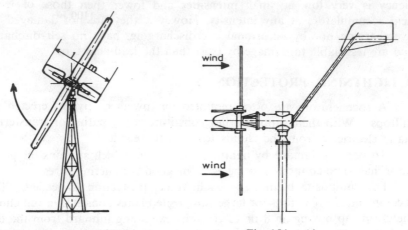

Fig. 153 – *Paris Rhône*.   Fig. 154 – *Aérowatt*.

rator.   The diameter of the area swept by the main blades varies between 2.40 m and 3.50 m according to the local wind speed.   The unit begins to produce energy when the rotational speed exceeds 325 r.p.m.

Aerowatt wind turbines (fig. 154)

The Aerowatt Company makes wind turbines from 24 watt to 4 100 watt capacity.

The rotor consists of two varible-pitch blades whose regulating device is controlled by centrifugal force acting on weights fastened to their axes. The blades are not twisted, and their chords are constant from the hub to the tip.   They are made of aluminium alloy (AG 3) and calculated to withstand windspeeds ranging from 56 m/s to 90 m/s.

The generators are generally permanent magnet models.   The most powerful units drive the generators through step-up gearings but the smaller wind turbines are directly coupled.

These wind plants are designed for rated wind speeds of 5 m/s or 7 m/s. The cut-in speed is about 3 m/s.   All of them are oriented by means of tail vanes.

In France, many lighthouses use Aerowatt wind turbines as a source of energy.   One example is the Seven Islands' lighthouse near Perros-Guirec in Brittany whose generator provides 5 kW power at a rotational speed of 300 r.p.m.   In that location, the wind turbine can be subjected to wind speeds of 60 m/s.   With a rotor diameter of 9.20 m, it supplies about 20 000 kWh/year.

Table 20 gives the main data relative to the aerogenerators manufactured by the Aerowatt Company.

TABLE 20

Model	24 FP	150 FP	200 FP	300 FP	1100 FP	4100 F
Diameter, m	1	2	3.2	3.2	5	9.2
Power, W	24	140	200	350	1 125	4 100
V_N, m/s	7	7	5	7	7	7
N, r.p.m.	1 200	525	380	420	178	142

Enag aerogenerators (fig. 155)

The ENAG firm produces sturdy aerogenerators with variable-pitched blades made of aluminium alloy. They are oriented by tail vanes. The generators are directly driven by the wind rotors. The most commonly used are direct current generators. However some wind plants are equipped with alternators. For example, the 5 kW aerogenerator is composed of a three-bladed rotor directly coupled to an alternator with 16 poles. The operating voltage is reached at a speed of about 120 r.p.m. The rated power is obtained when the rotational speed reaches 280 r.p.m. Table 21 gives the main specifications for the different models.

TABLE 21

Model	Two Blades		Three Blades	
Diameter, m	2.35	2.55	4.4	6
Power, W	650	1 000	3 000	5 000
V_N, m/s	9	9	9	11.5
Cut-in-speed	4 m/s	4 m/s	4 m/s	4.5 m/s

The energy is stored, as for Aerowatt plants, in lead acid batteries with thick plates and a capacity of at least 250-350 Ah.

Aerogenerator of the French firm Aéroturbine (fig. 156)

It consists of a variable pitch three-bladed wind rotor which rotates downwind of a support. The blades made of extruded aluminium alloy are not twisted and their chord is constant from the hub to the tip. The blade section is a NACA 64_4225 aerofoil. The generator (a brushless Alsthom alternator of 10 kW capacity) is driven by the wind rotor through a step-up gearing (gear-up ratio $k = 14.3$).

The regulating system enables the rotor to start very easily (high pitch angle for starting) like the Aerowatt regulator, and to feather when the

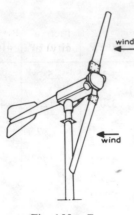

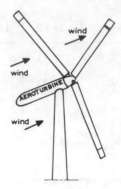

Fig. 155 – *Enag*   Fig. 156 – *Aéroturbine*.

wind velocity is too high, in the same way as the ENAG regulator. It
includes an electronic servomotor.

The machine is self-orienting. Its main characteristics are as follows :
Diameter : 8 m
Rated Power : 10 kW at 10 m/s windspeed
Rotational speed of the wind rotor : 105 r.p.m.
Rotational speed of the alternator : 1 500 r.p.m.

2. AMERICAN WIND TURBINES

Many firms in the USA are interested in wind turbine fabrication.
Table 22 gives the main specifications of some models.

TABLE 22

Trade mark	Windcharger	Sencenbaugh	North Wind	W.T.G.	Grumman
Model	20 . 110	500 .1 000	HR2	MP20	Windstream 33
Blades	2, wood. Fixed blades	3, wood. Fixed blades	3, wood. Fixed blades	3, stainless steel Fixed blades	3, alum. alloy. Variable Pitch
Rotor loc.	upwind	upwind	upwind	upwind	downwind
Diameter, m	2.30 - 3.35	1.80 - 3.60	5	8.5	10
Power, kW	0.25 - 1	0.5 - 1	2.2	20 (Max 36)	15 (Max 20)
V_N, m/s	8.5 9	11 10	9	13.5	10.8
N, r.p.m.	400	1 000 290	250	120	74.1
Gear-up ratio	1 1	1 . 3	1	15	25,1
Generator	d.c. . d.c.	d.c. . d.c.	d.c.	induction	induction
Orientation drive	Tail vane	Tail vane	Tail vane	Tail vane	Hydraulic servomotor

The last four wind generators are shown in figures 157a, 157b, 158a, 158b. The Windcharger unit and the Jacobs unit (which is also made in the USA in two models of 2 500 and 3 000 watt capacity) are described in chapter III.

The W.T.G. MP 20 and the Windstream 33 units are designed to operate coupled to the grid. The MP 20 wind generator uses the stalling effect of its fixed twisted blades above the rated wind velocity to limit power output.

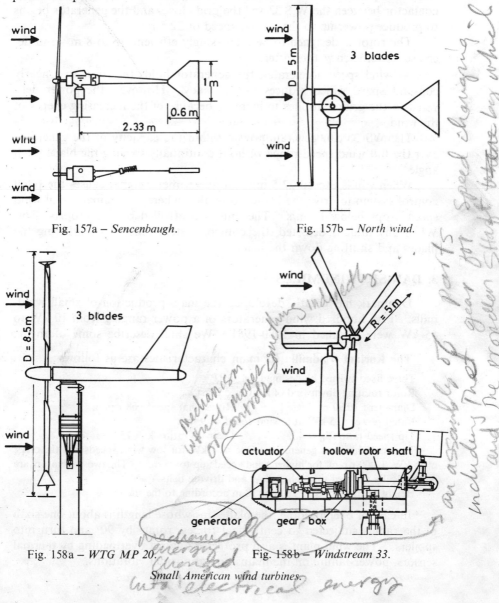

Fig. 157a – *Sencenbaugh.*   Fig. 157b – *North wind.*

Fig. 158a – *WTG MP 20.*   Fig. 158b – *Windstream 33.*

Small American wind turbines.

In case of overspeeding, the spoilers located at the blade tips, are deployed causing the rotor to slow down.   A disc-brake mounted on the high speed shaft is used for "parking" the rotor.

The Windstream 33 is designed to operate unattended.

When the power switch in the control panel is in the proper position, the blade pitch changes from the feathered to an intermediate position, causing the rotor to start turning.   When the wind speed reaches 4 m/s, the primary actuator drives the blades to high-speed running position, the contactor between the W.S.33 and the grid closes and the generator begins to produce power up to the furling speed of 22.5 m/s.

The rotor is designed to be increasingly efficient up to 8 m/s and decreases in efficiency thereafter.

As wind speed accelerates, the generator holds the rotor to a nearly constant speed which reduces its efficiency.   However, the power delivered to the grid continues to increase because of the increasing energy of the wind.

The WS 33 can generate power within the capacity of the generator over the full wind speed range without continually varying the blade pitch angle.

When winds exceed 22,5 m/s, an anemometer signal causes the pitch control system to drive the blades to feather where they remain until wind speed drops below 12 m/s.   The unit is controlled by a microprocessor. When any fault is detected, the control system reacts by feathering the blades and shutting down the machine.

3. DANISH WINDMILLS

Denmark has greatly developed the mass-production of small windmills.   Five hundred wind generators of a power ranging from 10 kW to 55 kW were installed in 1980-1981.   We shall describe some of them :

The Kuriant windmill : Its main characteristics are as follows :

Three fixed twisted blades made of GRP
Rotor rotating downwind of a latticed mast.
Diameter : 10.90 m   Rotational speed : 68 r.p.m.
Rated power : 15 kW at a wind speed of 9 m/s.
Tip speed ratio : $\lambda_0 = 4.2$ Gear-up ratio : k = 15
Two asynchronous generators : One of 4 kW for low wind speeds and the other of 15 kW, 1 000 r.p.m. for higher wind speeds up to 25 m/s.   The two generators are mechanically coupled through pulleys and driving belts.
Height of the support : 12 m or 18 m according to the site.

In case of overspeeding, the blade tips whose length is about one-sixth of the radius subjected to centrifugal forces, rotate by 90° and turn into spoilers.   The wind plant is also protected against overloading by thermal sensors, power-failure on the mains and abnormal vibrations.

A compressor placed at the foot of the guyed support provides compressed air which holds open the mechanical brake. In case of emergency, the pressure of the compressed air falls to zero, causing the disconnection of the generator from the grid and the mechanical braking of the wind rotor shaft which stops quickly.

The wind machine is oriented into the wind by a yawing motor controlled by a vane placed under the generator.

The Blacksmiths windmill drives an asynchronous generator of 22 kW capacity. The rotor diameter is 10 m. The three-fixed blades are made of fibreglass-reinforced plastic. The blade tips, controlled by centrifugal force, turn 90° at about 5 % overspeed and automatically turn back again when speed drops to 30 % of its normal value.

The Holger Danske windmill is similar to the previous one, except that its diameter is 11 m and the rotation of the blade tips is 60° at overspeed.

The Herborg windmill, largely similar to the Blacksmiths windmill, is equipped with two asynchronous generators of 30 and 5 kW capacities.

The Sonebjerg windmills are manufactured with asynchronous generators from 22 to 55 kW capacity and rotor diameters ranging from 10 to 14 m. The 3 blades are of fixed pitch and are equipped with spoilers operated by centrifugal force. Each blade consists of a wooden spar covered by a layer of fibreglass-reinforced polyester.

The Jydsk Vindkraft windmill drives an asynchronous generator of 15 kW capacity. The pitch of its 8.4 m fibreglass-reinforced polyester rotor is electrically controlled.

The **Windmatic firm** manufactures windmills having diameters of 10, 12 and 14 m and powers ranging from 10 kW to 55 kW.

The Windmatic wind turbines have three fixed blades connected together with stays. Each blade consists of a main supporting beam of laminated wood around which the fibreglass profiles with the boxes of the braking flaps are moulded.

The main specifications of the 10 m-22 kW Windmatic machine are as follows :

Three fixed blades made of GRP running upwind of the support
Diameter : 10 m   Rotational speed : 68 r.p.m.
Power delivered : 10 kW in a 8 m/s wind, 22 kW in a 12 m/s wind.
Tip-speed ratio : $\lambda_0 = 4.4$ Gear-up ratio : k : 14.83
Cut-in-speed : 5 m/s   Furling speed : 20 m/s
Generator : 22 kW, 1 000 r.p.m.   Lattice-steel mast, 18 m high.

The rotor is oriented into the wind by two lateral auxiliary wind wheels. Each unit is equipped with safety devices. They protect the machine against overloading, overspeeding, and vibrations. In case of overspeeding which can happen if the machine is disconnected from the grid, the spoiler flaps fan out. A mechanical brake allows the machine to be stopped.

The Erini windmill is similar to the above, except that the blades are of welded construction. The blade tips turn 90° at overspeed under the influence of centrifugal force.

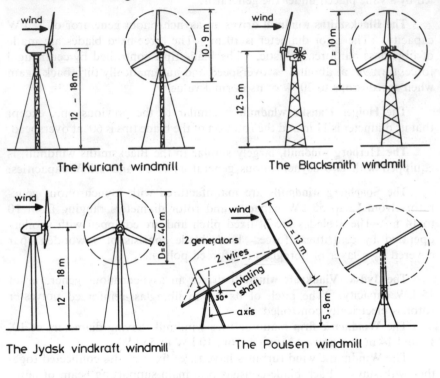

The Kuriant windmill   The Blacksmith windmill

The Jydsk vindkraft windmill   The Poulsen windmill

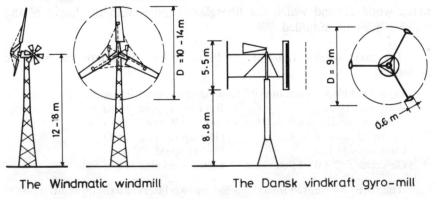

The Windmatic windmill   The Dansk vindkraft gyro-mill

Fig. 159 – *Danish windmills*

The Poulsen windmill is an unusual one.   The shaft is inclined 30°
to the horizontal plane.   The fixed pitch, two-bladed rotor drives a 30 kW,
4 pole asynchronous generator at 120 r.p.m. in strong winds, and a 5 kW,
6 pole generator at 60 r.p.m. in light winds.

The Dansk Vindkraft gyromill drives an asynchronous generator of
15 kW capacity.   The blade pitch angle is controlled by means of a vane.
    Note that the Danish manufacturers are in favour of the induction
generator because this can be connected to the network without difficulty.

Experimental Danish wind power plants

    Before undertaking the construction of the Gedser windmill, J. Juul
and his collaborators built two other smaller experimental windmills at
Bogo and Vester.

    We shall describe only the Vester wind plant.   Figure 160 gives the
main data concerning its blades and figure 161 shows its regulating

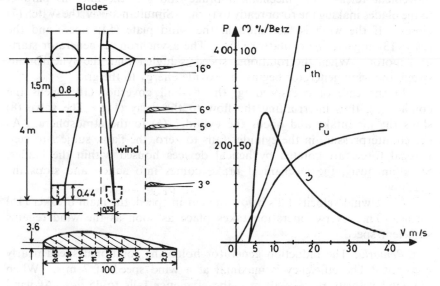

Fig. 160 – *Danish Vester windmill.*

    We shall note : (1) the blades, (2) the ailerons, or flaps at the blade tips,
(3) the mechanical brake, (4) the gear box, (5) the generator, (17) the com-
pressed-air or oil intake.

    The windmill is controlled by means of a mobile rod operating through
a hollow windmill shaft and welded to a piston (6) acting on two sliding
rods connected to the flaps.

    When the circuit breaker (16) and the switch (12) are closed, the
relay (11) of the electromagnetic valve (8) is fed.   Compressed air going

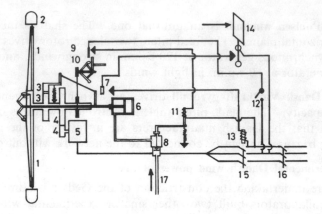

Fig. 161 – *Regulating device.*

through the open valve (8) causes the piston (6) to move to the right.   This movement releases the mechanical brake and sets the ailerons parallel to the blades making the rotor ready to turn.   Simultaneously, the switch (7) closes.   If the wind is fast enough, the wind plate (14) closes and the relay (13) engages the contactor (15).   The asynchronous generator starts as a motor.   When the rotational speed is higher than the synchronous speed, the wind generator begins to provide energy to the grid.

In the case of overspeeding, the fly-ball governor (10) opens the contact (9), thus interrupting the flow to the relay (11).   The valve (8) shuts the air intake and opens the cylinder (6) to the atmosphere.   As the counterpressure in the cylinder falls to zero, the flaps, subject to centrifugal force, are rotated by helical devices housed within the blades. Simultaneously, the mechanical brake comes into action and stops the machine.

If the wind velocity falls below the cut-in speed, the main breaker (15) opens.   The reverse operation takes place as soon as the wind regains sufficient speed.

Remark : The induction generator holds the rotational speed nearly constant.   The efficiency is maximal at a wind speed of 7 m/s.   When the wind velocity reaches 10 m/s, the efficiency falls to 45 %.   At wind speeds over 10 m/s, the efficiency is so poor that a speed-regulator would be unnecessary to maintain the rotational speed.   Therefore the aerodynamical brake is useful only when the wind generator is accidentally disconnected from the grid.

4. SOVIET WIND MACHINES

Table 23 gives the main data of some Soviet aerogenerators.   These machines are described in *Wind Powered Machines,* a Soviet book written by Y. I. Shefter and translated into English by NASA.

TABLE 23

Model	BE 2M	VIESKH 4	UVEU D6	SOKOL D12
Blades	2	2	2	3
Diameter, m	2	4	6	12
Power, W	0.15	1.6	3.4	15.2
V_N, m/s	6	8	8	8
N, r.p.m.	600	280	186	88
Productive wind speeds, m/s	3-25	4-40	4-40	4.5-40

The first three of these machines are equipped with the Soviet regulator seen in fig. 71, Chapter III. The Sokol wind machine is provided with a centrifugal spring regulator.

5. SWISS WIND MACHINES

The Elektro firm of Winterthur built aerogenerators ranging from 50 W to 6 000 W capacity. The regulating system is described in Chapter III. The generators currently in use are direct-driven alternators. Some of them are equipped with rectifiers for battery charging. Table 25 gives the main data. An Elektro wind turbine is shown in fig. 163.

TABLE 24

Model	W 50	W 250	WV 05	WV 15	WV 25	WV 35	WV 50
Blades	1	1.3	2	2	2	3	3
Diameter, m	0.45	0.66	2.5	3	3.6	3.4	5
Power, W	50	250	600	1 200	2 200	4 000	6 000
N, r.p.m.	100/450	70/400	250/700	220/550	200/520	160/420	120/220

6. AUSTRALIAN WIND TURBINES (fig. 164)

The Dunlite firm of Adelaide has built two models of aerogenerator which can respectively provide 1 kW in a 10 m/s wind, and 2 kW in a 12 m/s wind. The three-bladed wind rotor, which are identical for the two models, have diameters of 4,10 m. The blades are made of steel. The current produced by both alternators is rectified. The variable blade pitch is controlled by a regulating system of the Quirk type.

7. GERMAN WIND TURBINES

We shall describe some of the models which are tested on Pellworm Island where the German wind energy test field is located.

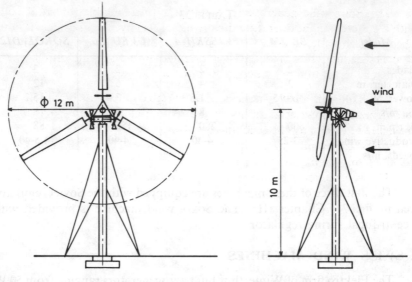

Fig. 162 – *The Sokol wind generator* (U.S.S.R.).

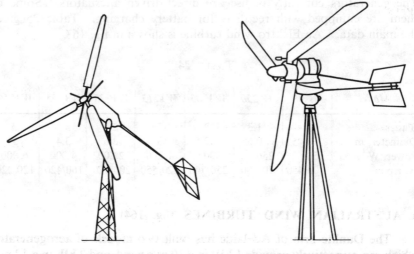

Fig. 163 – *Elektro wind rotor*   Fig. 164 – *Dunlite wind turbine*
(Swiss)   (Australia).

The Brummer wind generator is a variable-pitch, three-bladed wind machine operating downwind of its guyed steel-tubed mast. The blades are made of aluminium and the machine is self-orienting.

The Aeroman wind generator is a variable-pitch two-bladed wind rotor oriented by an auxiliary wind wheel and rotating downwind of its support. The blades are made of fibreglass-reinforced polyester (GRP).

The Böwe wind generator is a self-orienting, one-bladed wind rotor with a counterweight operating downwind of its guyed cylindrical support. The blade is made of GRP. The speed is controlled by a slat. In case of overspeeding, the blade axis is oriented into the wind direction by means of a hydraulic jack actuated automatically.

The Hüllman wind generator is a variable-pitch three-bladed wind rotor rotating downwind of its guyed cylindrical support. The blades are made of GRP and the machine is self-orienting.

All these wind rotors drive synchronous generators (220/380 V, 1 500 r.p.m.) having more than 10 kW capacity.

The geometrical and mechanical characteristics are given in table 25.

TABLE 25

Model	Brümmer	Aeroman	Böwe	Hüllman
Diameter	12 m	11 m	12 m	9 m
Number of blades	3	2	1	3
Rated power	10 kW	11 kW	10 kW	10 kW
Rated wind speed	8 m/s	8 m/s	8 m/s	9 m/s
Rotational speed	40 r.p.m.	100 r.p.m.	115 r.p.m.	100 r.p.m.
Tip-speed ratio λ_0	3.14	7.2	9	5.9
Gear-up ratio k	37.5	15	13	15
Height of the support	9 m	10 m	11 m	12 m

Another model : Noah contra-rotating wind rotors

In order to avoid the need for step-up gearing, W. Schoenball has built a 12 m diameter wind turbine with two five-bladed rotors, one rotating clockwise and the other counter-clockwise. One rotor turns the rotor of the generator, the other turns its stator. Because rotational speeds of the wind rotors are opposite, step-up gearing is unnecessary. The speed of each rotor is maintained at 71 r.p.m. by means of an electrical device. An auxiliary wind wheel turns the main rotor into the wind. The prototype built on Sylt Island in Germany has supplied a power of 70 kW. The advantage in the layout of avoiding step-up gearing is unfortunately offset by its greater complexity, which constitutes a serious handicap for this machine.

8. BRITISH WIND POWER PLANTS

During the last decades, four experimental wind power units were constructed in the United Kingdom.

Enfield wind turbine : This two-bladed wind rotor was coupled directly to a three-phase low-revolution generator operating at 415 v.

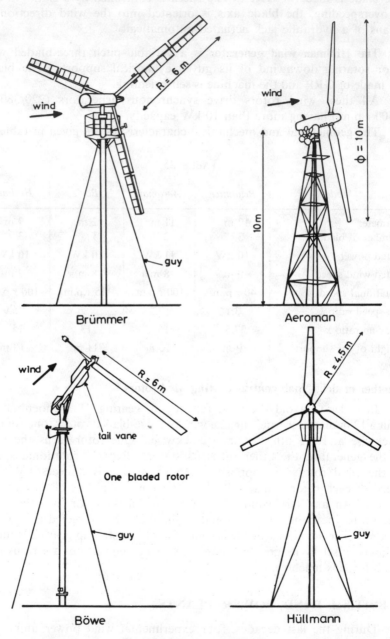

Fig. 165 – *German windmills.*

Dowsett wind machine : This machine was a three-bladed wind turbine with a hydraulically-controlled variable pitch which drove a 25 kW induction generator.   It was oriented by two auxiliary rotors.

Smith wind turbine (Isle of Man): The machine had three fixed blades with constant chord, made of extruded aluminium.   It was coupled to a 100 kW induction generator and was oriented by an auxiliary wind wheel.

John Brown wind turbine (Costa Hill, Orkney): This installation included a variable-pitch three-bladed wind rotor which drove a 100 kW induction generator.   The blades were tapered and untwisted.

Table 26 gives their main specifications.

TABLE 26

Wind machine	Enfield	Dowsett	Smith	J. Brown
Blades	2	3	3	3
Diameter, m	10	12.8	15.2	15
Power, kW	10	25	100	100
V_N, m/s	8.3	11	18.5	15.2
N, r.p.m.	103	65	75	130
Tower Height, m	12	10	10.5	12

C. High-Power Installations

Most of the high-power systems constructed so far have been equipped with two or three-bladed turbines of fixed or variable pitch. Speed is regulated by aerodynamic brakes or blade adjustments, generally with the aid of servomotors because of the magnitude of the effort required.

For orienting the wind rotor into the wind, a wide range of devices may be used : yawing motors, auxiliary wind wheels, self-orientation etc...

Step-up gearings are generally inserted between turbines and generators. This permits avoiding the use of multipole generators, which are very heavy and necessary for a direct drive by the turbine shaft, due to the relatively low rotational speed of large wind rotors.   Step-up gearings mainly consist of multistage gear-boxes and gearing belts (efficiency 98 % or 99 % per stage).   Hydraulic transmissions are less common, though their performance has been improved in recent years.

In the following paragraphs, detailed information is given on the main high-power installations.

1. ANDREAU — ENFIELD AEROGENERATOR (fig. 166)

This aerogenerator was designed by the French engineer Andreau and manufactured by the English Enfield Company for the British Electricity Authority.

Initially, the machine was erected at St Albans (U.K.). Unfortunately, this was a wooded site where the wind was irregular, so the experiment was inconclusive. In 1957, the machine was bought by " Electricité et Gaz d'Algérie", dismantled, and then reerected at Grand Vent (Algeria).

Overall view   Cross section

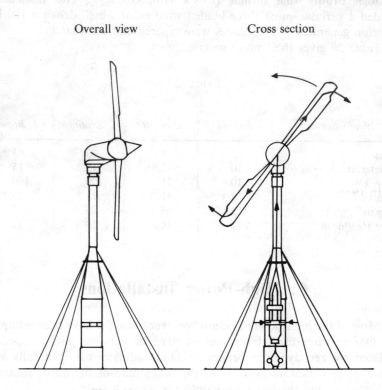

Fig. 166 – *Andreau-Enfield wind machine.*

Description

The design was unique. The blades were hollow. When they rotated, centrifugal force caused the air inside the blades to move from the hub towards the tips. The low pressure within the hub was used to drive an air turbine and an alternator, both placed inside the tower.

Main characteristics

Two articulated, hollow blades. Self-orienting wind turbine, assisted by a

sensitive power control system.   Automatic pitch control by a hydraulic servomotor.
Variable coning.

> Rated power : 100 kW   Rated wind speed : 13,5 m/s
> Power output is held constant from 13,5 m/s to 29 m/s windspeed.
> Rotational speed of the turbine : variable (maximum : 95.4 r.p.*m*.)
> Air-sucked discharge : 1 655 m³/min.
> Alternator : 100 kW, 415 V.
> Tower height : 30 m.

Results

Low efficiency (22 %) due to air intake at the rotating joints near the
hub, and also due to the fact that the efficiencies of the four elements placed
in succession (wind turbine, fan, air turbine, alternator) must be multiplied
together to obtain the overall efficiency.   Though the blades were of
necessity not perfectly streamlined, their efficiency reached 73 % relative
to the Betz limit.

The main conclusion was that it would be advisable to avoid articulated
blades.

2. AMERICAN WIND MACHINE (MOD-O)

In 1975, the U.S. Department of Energy erected a 100 kW wind turbine in Sanduski, Ohio.   Its characteristics are as follows :

> Wind rotor operating downwind of its support.
> Two variable-pitched blades made of aluminium.
> Diameter : 37.5 m.
> Rated power : 100 kW   Rated wind speed : 8 m/s
> Rotational speed : 40 r.p.m. in wind > 3 m/s by changing the system load and
> the pitch of the blades.
> Safety device : the blades are fully feathered in wind speeds greater than 30 m/s
> and the system is designed to withstand wind velocity as high as 70 m/s.
> Alternator : 125 kVA at 1 800 r.p.m. for MOD-0.
> Gear-box ratio : k = 45   Tower height : 30 m.

Four other installations called MOD-OA with the same geometrical
characteristics as MOD-O, but using 200 kW alternators, have been erected.
One of them was installed at Dayton, New Mexico, in December 1977
(Rated wind speed 11 m/s).

3. BALAKLAVA RUSSIAN WIND MACHINE (USSR)

This wind machine was erected in 1931 and operated until 1949.   Its
annual output was about 200 000 kWh.

Main characteristics

> Three-bladed wind rotor placed upwind of the latticed tower.
> Variable pitch controlled by flaps.

Diameter : 30 m.
Rated power   : 100 kW at 10.5 m/s wind speed.
Maximum power : 130 kW at 11 m/s wind speed.
Rotational speed : 30 r.p.m.   Tip-speed ratio : $\lambda_0 = 4.5$.
Induction generator : 220 V   Tower height : 25 m.

Orientation by means of an inclined strut, whose base rested on a circular track and was moved by a 1 kW electric motor.

The main gears were of wood and the blade skins of roofing metal.

The Zwei D-30 aerogenerator was a similar wind machine; the only difference concerned the yawing system. For orienting this machine, auxiliary wind wheels were used, the wind rotor being placed downwind of the tower.

Soviet aerogenerators

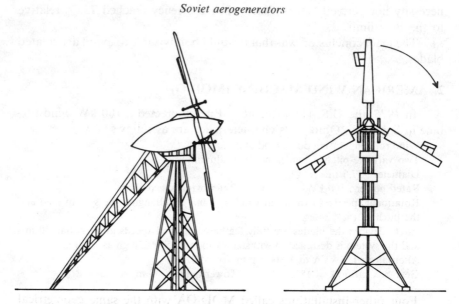

Fig. 167 – *The Balaklava machine.*   Fig. 168 – *The ZWEI D30 aero-generator.*

4. THE 132 kW NEYRPIC AEROGENERATOR, ST-REMY DES LANDES (FRANCE)

This wind machine (shown in fig. 169) was designed by Louis Vadot. It supplied 700.000 kWh between November 1962 and March 1966.

At first it was oriented by auxiliary wind wheels but tests showed that self-orientation was quite satisfactory, the wind rotor being placed in both cases downwind of the tower.

Main data

Three variable-pitched blades of fibreglass reinforced plastic.

Diameter : 21.1 m   Rated power : 132 kW.
Rated wind speed : 12.5 m/s   Rotational speed : 56 r.p.m.
Maximum power : 150 kW for 13.5 m/s wind speed.
Induction generator rotating at 1 530 r.p.m.
Double step-up gearing.
The power and efficiency curves are seen in fig. 169.

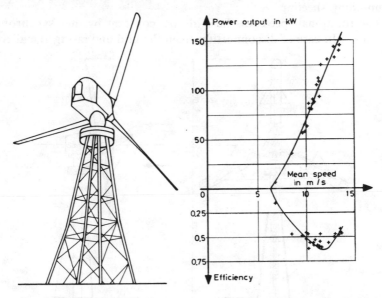

Fig. 169 – *The 132 kW Neyrpic aerogenerator at St-Rémy-des-Landes*
(Manche, FRANCE)

5. GEDSER DANISH WINDMILL (fig. 170)

This aerogenerator, designed by J. Juul, operated from 1957 to 1966.
On the average, it provided about 450 000 kWh/year, which corresponds
to 900 kWh/year/m² of swept area.

Main specifications :

Wind rotor rotating upwind of the tower.
Three fixed twisted blades. Profile : NACA 4312.
Useful blade length : 9 m   Blade chord : 1.54 m.
Setting angle near the hub : 16°, at the tip : 3°.
Diameter : 24 m   Rotational speed : 30 r.p.m.
Power : 60 kW at 7.5 m/s, 200 kW at 17 m/s wind speed.
Cut-in speed : 5 m/s   Cut-out : 20 m/s.
Design tip-speed ratio : $\lambda_0 = 5$.

Induction generator : 200 kW, 8 poles, slip : 1 % at full load.
Gear-up ratio : k = 25 (double chain).
Concrete tower, 24 m high.

In order to reduce the bending stresses at the hub, the blades were
connected together by stays.   The blade construction was derived from the
techniques used classic windmills.   The blades consisted of wooden
frames fixed on steel beams.   The wooden frames were covered with alu-
minium-alloy sheeting.

The rotational speed was held almost constant by the asynchronous
generator.   In case of disconnecting from the grid and racing, the ailerons

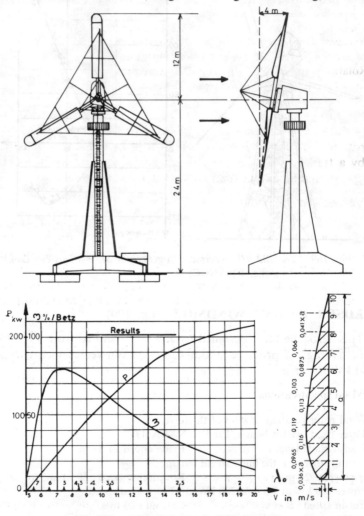

Fig. 170 – *GEDSER windmill* (Denmark).

placed at the blade tips rotated 60° by the action of a servomotor controlled by a flywheel regulator.   The aileron area was equal to 12 % of the blade surface.   The wind machine was stopped by a mechanical brake.

6. HUNGARIAN WIND TURBINE   (fig. 171)

In 1960, Ledacs Kiss and a team of technicians designed an experimental wind machine.   The installation shown in fig. 171 had the following specifications :

Wind rotor placed upwind of a concrete tower.
Four braced fixed twisted blades, 2 800 kg each, made of a welded steel structure• with an external aluminium covering.

Diameter : 36.6 m	Rated power : 280 kW
Rated wind speed : 10.4 m/s	Cut-in-speed : 3 m/s
Maximum power : 280 kW at 12 m/s wind-speed	
Rotational speed : 17.85 r.p.m.	Tip-speed ratio : $\lambda_0 = 5$
Induction generator : 1 000 r.p.m.	Tower height : 36 m

Step-up gearing: The shaft rotated on roller bearings.   At the rear end of the shaft was a sprocketed wheel with a chain drive ratio of 4 : 1. The rotation, having been multiplied by four by the chain-drive, was carried by a transmission shaft to a speed-increaser gear which had a ratio of 14 : 1, so that its output shaft rotated at 1 000 r.p.m.

Yawing and safety devices :

During periods of non-productive winds, the generator was disconnected from the grid by an automatic device.   In normal running, the nacelle was only turned away from its direction by a yaw motor when the wind direction had already changed by at least 15°.   The return rotation of the nacelle had the same amount of dead space, so that the wind wheel and nacelle did not turn back and forth at every slight change in wind direction.

Limitation of power output during strong winds was performed by an automatic device which turned the wind rotor a little out of the wind. If the wind velocity later decreased, the same automatic device turned the rotor back into the wind.   The wind turbine was not equipped with an aerodynamic brake, only with a mechanical one.   If need arose of complete shutdown, the plane of the wind wheel could set completely parallel to the wind direction.   Then the rotor was stopped by the mechanical brake.

7. AEROGENERATOR AT PETTEN, NETHERLANDS (fig. 172)

In 1980, the Dutch Department of Energy installed an aerogenerator near Petten under the leadership of G. Piepers.

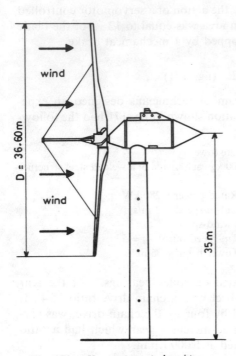

Fig. 171 – *Hungarian wind turbine.*

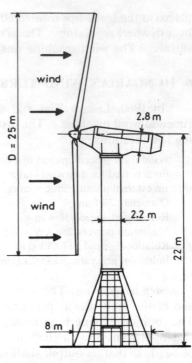

Fig. 172 – *Dutch aerogenerator of Petten.*

Its main specifications are the following:

Two-bladed wind rotor running upwind of the tower.
Fibreglass construction, beam section reinforced by carbon fibre.
Variable pitch controlled by a hydraulic actuator.

Aerofoil : NACA 2300	Twist-angle : $16°$
Cone angle : $5°$	Tilt angle : $5°$
Diameter : 25 m	Blade weight : 550 kg each
Power : 400 kW	Rated wind speed : 13 m/s
Cut-in speed : 6 m/s	Cut-out speed : 17 m/s
Design rotational speed 80 r.p.m.	Design tip-speed ratio : $\lambda_0 = 8$
Rotational speed range : 40-80 r.p.m.	
Planetary gear-box : $k = 20$	Yaw rate : 1.1 deg/s

d.c. Shunt generator : 400 kW, 900 1 600 r.p.m.
6-phase bridge converter with line commutation, 600 kW.
Step-up transformer : 380 V to 10 kV.
Concrete tower with a conical base

The speed of rotation N is variable and selected by computer in order to extract the maximal power for a given wind velocity.

The d.c. current provided by the shunt generator is transformed into a.c. current (180 V, 50 Hz) by means of a converter.   This system is well-suited for transferring power from a large wind rotor to a "weak" grid. It suppresses the problems of stability caused by gusts by allowing the rotor to accelerate.

Note that the Dutch engineer L. Lievense has studied the possibilities of storing wind energy by linking 1 000 wind turbines of 3 MW each to a massive water basin of 165 km2.   The wind would provide power to pump water into the basin at a level higher than that of the surrounding Ijsselmeer. This water would then be allowed to flow out, driving water turbines to supply power when needed during periods of non-productive winds.

8. BEST ROMANI AEROGENERATOR AT NOGENT-LE ROI, FRANCE (fig. 173)

This aerogenerator was designed by Louis Vadot.   Between April 1958 and April 1962, it supplied 221 000 kWh to the grid.

Its main characteristics were as follows :

Self-orienting wind rotor rotating downwind of its support.
Three fixed twisted blades made of aluminium alloy.
Profiles used : NACA 23012, 23015, 23018
Diameter : 30.1 m   Rated power : 800 kW
Rated wind speed : 16.7 m/s   Rotational speed : 47.3 r.p.m.
Tip-speed ratio : $\lambda_0 = 7$ Gear-up ratio : k = 21.5
Generator : Alternator with 6 poles running at 1 000 r.p.m.
Cut-in-speed : 7 m/s   Height of pylon : 32 m.
Total weight without infrastructure : 160 t.
The wind rotor was designed to rotate satisfactorily in winds of 25 m/s with gusts up to 35 m/s and to withstand 65 m/s wind speeds when stopped.

Power was transmitted from the wind rotor to the generator by two planetary gear-boxes with ratios of 7.5 : 1 and 3 : 1 respectively.

The effect of the support wake was reduced to a minimum by means of slots sucking air between the circular section of the mast and a moving hollow beam in the shape of an aerofoil tail, welded to the body of the wind machine.   This also carried a ladder for climbing into the nacelle.

For starting, a clutch allowed the wind rotor to run freely.   The alternator started like an asynchronous motor.

In case of disconnection from the grid and racing, an automatic device established an electrical connection between the alternator and an auxiliary network consisting of a resistive line 60 m long.   This electrical brake, and a disc brake 1.80 m in diameter, stopped the machine within two revolutions of the wind rotor shaft.

The aerogenerator was used for five years as an experimental platform. During a storm, it provided 1 000 kWh for about twelve hours.   Its effi-

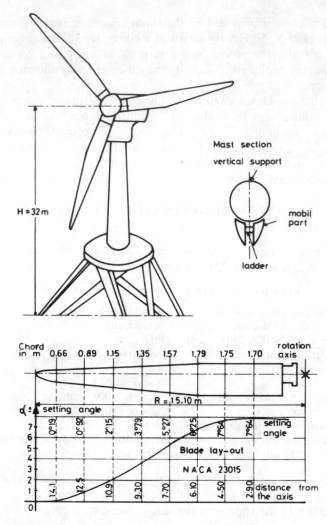

Fig. 173 – *The 800 kW Best Romani aerogenerator* (France).

ciency reached 80 % of the Betz limit at optimal speed.   During tests,
on August 30, 1960, the output power passed from 300 kW to 900 kW
within an interval of 2.85 seconds.

For reducing the step-up gearing ratio, a second wind rotor rotating
at 71 r.p.m. with a corresponding tip-speed of 112 m/s was tested in 1963,
but one of its blades broke.   L. Romani, Head of the Eiffel Laboratory
and Chief manager and inventor of the aerogenerator, estimates 100 m/s
as the safe tip-speed limit.

After the blade broke, the machine was dismantled and left unrepaired
because of low petroleum prices at the time.

9. NEYRPIC AEROGENERATOR OF 1 000 kW POWER, ST-REMY DES LANDES, MANCHE, FRANCE (Fig. 174)

This aerogenerator was designed by Louis Vadot and operated from 1963-1964.

Its operation, which was quite satisfactory, was stopped after 2 000 hours because of damage to the step-up gearing. In spite of the size of the investment and the short duration of the run, the wind machine was not repaired because of the then low petroleum prices. The aerogenerator was dismantled in June 1966.

Main characteristics

Variable-pitch three-bladed machine controlled by a hydraulic servomotor. Blades made of fibreglass-reinforced plastic (G.R.P.).

Self-orienting wind rotor, downwind of the pylon.
Diameter   : 35 m
Rated power : 1 000 kW at 17 m/s wind speed.
Efficiency : 60 % to 70 % relative to Betz limit.
Height of the pylon : 61 m.
Fixed pitch up to 650 kW and variable pitch thereafter.
Stoppage by feathering for V < 6 m/s.
Asynchronous generator : 1 015 r.p.m., 3 000 V.
Total weight (excluding infrastructures) : 96 t.
The power and efficiency variation curves are given in fig. 174.

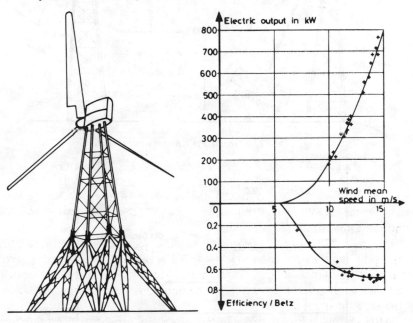

Fig. 174 – *The 1 000 kW Neyrpic aerogenerator at St-Rémy-des-Landes* (Manche, FRANCE).

10. GRANDPA'S KNOB WIND MACHINE (VERMONT, USA)
   (fig. 175)

This machine was built by the S. Morgan Smith Company under the leadership of P. C. Putnam and J. B. Wilbur. It ran from October 1941 to March 1945.

On March 26, 1945, a blade failed. The machine was not repaired because of the low petroleum price at the time. The company abandoned the project and placed its patents in the public domain. Until 1975, the Grandpa's Knob was the greatest wind machine ever erected.

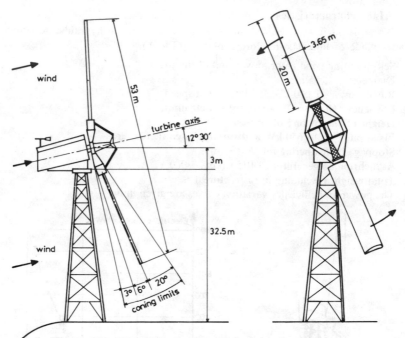

Fig. 175 – *The Grand Pa's Knob wind turbine.*

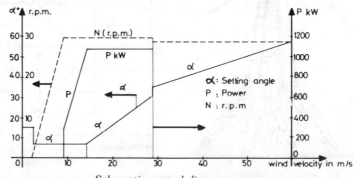

Schematic control diagram.

Its main characteristics were as follows :

Variable-pitch two-bladed wind rotor operating downwind of the tower. oriented by a hydraulic yaw motor.

Rectangular and untwisted blades made of stainless steel.

Profile used : NACA 4418   Constant chord : 3.70 m
Useful blade length : 20 m   Weight of each blade : 6.9 t
Diameter : 53 m   Rated power : 1 250 kW
Rated wind speed : 13.5 n   Rotational speed : 29 r.p.m.
Tilt angle 12°5, Cone angle : variable from 6° to 20°
Alternator : 1 250 kVA, 2 400 V, 600 r.p.m.   Gear-up ratio k = 20.6
Height of the lattice tower : 33 m   Total weight : 75 t

The regulation of the machine was carried out through a hydraulic servomotor acting on the blade pitch.

In order to reduce the bending stresses in the root sections of the blades, the angle between the rotor shaft and the blade axes varied with the wind velocity and the rotational speed.   Coning variation was possible.

The wind rotor was able to withstand 62 m/s wind speeds.

To avoid the alternator being forced out of phase and to hold the connection to the grid, the wind rotor drove the generator through a hydraulic coupling capable of slipping.   Power peaks from gusts were absorbed by allowing the rotor to accelerate.

11. DANISH WIND GENERATORS

Denmark has greatly developed the use of wind energy by constructing several wind generators of medium and large size.   The largest ones have been constructed in the localities of Tvind and Nibe.

a) AEROGENERATOR at TVIND

This wind turbine seen in fig. 176 was erected in 1977 by the inhabitants of Tvind, a village of 700 people situated on the west coast of Denmark, to supply their own energy needs.

Its main characteristics are the following:

Variable-pitch wind rotor rotating downwind of a tower.
Three-blades made of fibreglass-reinforced polyester (GRP).
Profiles used : NACA 23035, 23024, 23012   Weight : 3.5 tons each
Cone angle : 9°   Tilt angle : 4°
Diameter   : 54 m   Rated power : 2 000 kW
Rated wind speed : 15 m/s   Maximum wind speed : 20 m/s
Rotational speed : 40 r.p.m.   Tip-speed ratio : $\lambda_0 = 7,5$
Alternator : 2 000 kW, 3 kV, 750 r.p.m.
Gear-up ratio : k = 19   Gear-box weight : 18 tons
Concrete tower 53 m high.

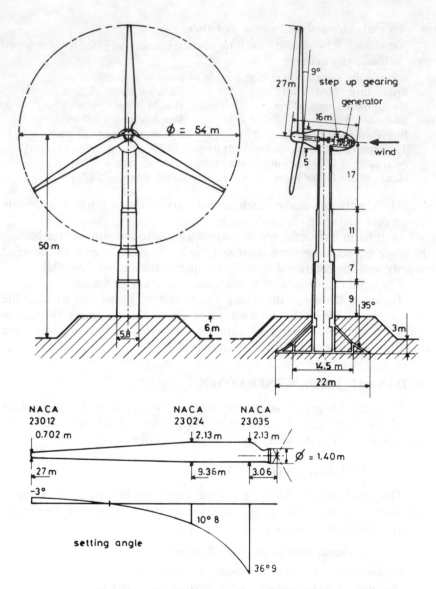

Fig. 176 — *The TVIND wind turbine* (Denmark).

The shaft from the rotor hub to the gear-box is a solid steel cylinder of 800 mm diameter. It originally served as a ship propeller shaft.

The electricity is fed down from the top of the tower by loose cables which are slack enough to require untwisting only about once a year. Before the electricity leaves the machinery pod, it goes through a solid-state rectifier. This thyristor device converts the varying-frequency a.c. current from the generator into d.c. and then back into a.c. at the desired frequency.   According to time of day, this frequency varies from 48 Hz to 50 Hz.

b) NIBE WIND TURBINES (fig. 177)

The Nibe site was equipped with two models of wind turbine in 1978 and 1979.   They are placed 150 m from the water's edge at a distance of only 220 m from each other.   The design of the wind rotors and the regulating systems are different for the two machines.   The aim of the experiment is to form the basis for future decisions concerning wind energy conversion systems for electricity production.

Main characteristics :

Turbine A : 3 blades, upwind rotor.
4 discrete pitch angles stall regulated.
Rotor blades supported by stays.
Turbine B : 3 blades, upwind rotor.
Full pitch control for power regulation.
Rotor blades self-supporting.

Characteristics common to both wind turbines :

Rotor diameter : 40 m   Height of the rotor hub : 45 m
Cone angle : 6°   Tilt angle : 6°
Blade construction : Steel/fibreglass spar, fibreglass shell
Profile NACA 4412-4434, Standard roughness, Twist angle : 11°
Rated power : 630 kW   Rated wind speed : about 13
Rotational speed : 34 r.p.m.   Tip-speed ratio : $\lambda_0 = 5.5$
Cut-in-speed : 6 m/s   Cut-out speed : 25 m/s
Induction generator, 4 poles, 6 kW, 1 500 r.p.m.
Gear-box : Three stage gear, ratio 45 : 1
Automatic control via computer
Hydraulic yaw motor, yaw rate : about 0.4 deg/s
Electricity lead down   : 3 single-phased turnable cables.
Estimated production   : 3 GWh per year
Maximum power produced: 500 W/m^2 swept area
Low-tuned concrete tower, 41 m high.
Weight of the outer 12 m blade section :   900 kg
Weight of a rotor blade (turbine B)   : 3 500 kg
Weight of a nacelle and rotor blades   : 80 tons.

These wind turbines (seen in fig. 177) have been studied by Helge Petersen and the engineers of the Riso National Laboratory.

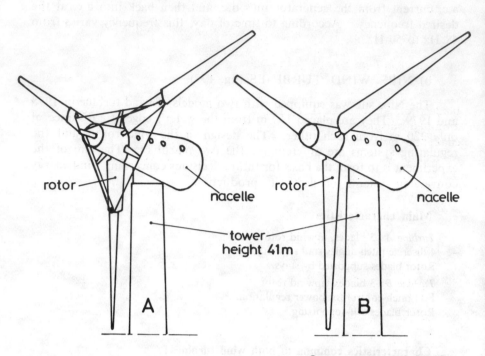

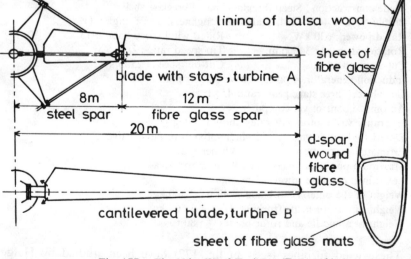

Fig. 177 – *The Nibe Wind Turbines* (Denmark).

c) THE 265 kW VOLUND WIND TURBINE (fig. 178)

This machine designed by Helge Petersen includes two induction generators : a 58 kW for low wind speeds and a 265 kW for maximum production in higher wind speeds.   Its efficiency is high over a large range of wind velocities.

The blades are mounted in such a way that the bending stresses in the blade roots are eliminated by employing a system of struts.   The hinges mounted outside of the blade structure and the inner bearings at the hub enable the blades to pitch approximately 100º.   The blades can turn under the action of the centrifugal and aerodynamic forces acting on them, thereby eliminating the need for active safety control devices.

The rotor is provided with a tail in the shape of a small multi-bladed windmill.   This assures the orientation of the machine into the wind. Thus there is no need for a yaw motor.

Its main characteristics are the following:

Three-bladed rotor operating downwind of its support with a rigid hub.   Electro-mechanically-controlled variable pitch.

Blades made of GRP, Profile NACA 4412-4420, Twist : 7.8º.

Root chord : 1.80 m	Tip chord : 0.60 m
Cone angle : 4º	Tilt angle : 6º
Diameter : 28.3 m	Rated power : 265 kW
Rated wind speed : 13 m/s	Cut-in speed : 4.5 m/s
Cut-out speed : 22 m/s	Maximum design speed : 70 ms

Rotational speed : 28 r.p.m. at low wind speeds
                              42 r.p.m. at high wind speeds
Transmission : Three-stage, ratio k = 36
Two induction generators directly coupled
Rating : 6 poles, 1 008 r.p.m., 58 kW, 380 V
                4 poles, 1 512 r.p.m., 265 kW, 380 V
Estimated annual output :
For $\overline{V}$ = 5.40 m/s at 10 m above the ground : 450,000 kWh
                6.30 m/s   630,000 kWh
                7.20 m/s   800,000 kWh
Nacelle made of wound fibreglass, length : 6 m, diameter : 2.4 m
Tower made of tubular steel with four guys
Diameter at the base : 1 400 mm, at the hub : 800 mm
Weight of the rotor including blades : 4 tons, above tower : 14 tons
Tower : 6 tons   Total weight : 20 tons

12. NEW AMERICAN WIND GENERATORS

Since 1975, the United States have been developing an ambitious program in the wind energy field.

This program began with the construction of the wind turbine MOD.0

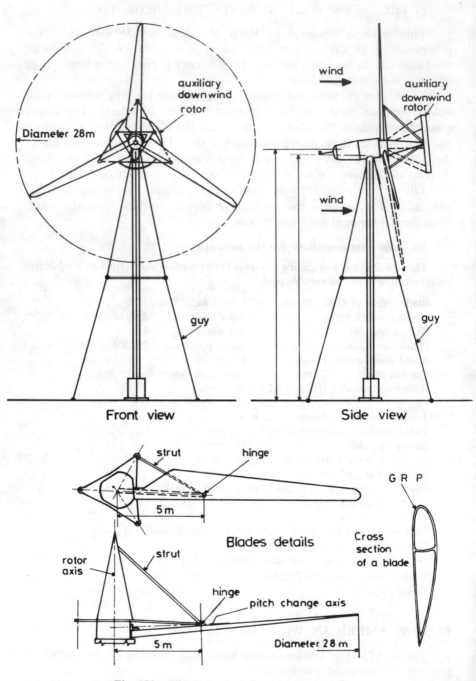

Fig. 178 – *The Volund wind turbine* (Denmark).

of 100 kW power, described in a previous paragraph. Then four other installations called MOD.0A, having identical wind rotors and shapes but equipped with generators of 200 kW power, were erected at different sites in the U.S.A.

In 1978 and 1980, the U.S. Department of Energy, under the leadership of L. V. Divone, Director of the Wind Energy Systems Division, constructed two large wind generators MOD.1 and MOD.2.

a) THE MOD.1 WIND TURBINE (fig. 179)

The MOD.1 American wind turbine (2 000 kW) has been operating at Boone, North Carolina, at the summit of Howard's Knob since the summer of 1979. It was manufactured by NASA, General Electric and Boeing.

Its main characteristics are the following:
Variable pitch, two - bladed wind rotor rotating downwind of a truss tower with a rigid hub.

Blades made of stainless steel   Profil NACA 44XX
Linear chord (at root : 3.65 m, at tip : 0.85 m), twist angle : $11°$

Cone angle : $9°$	Tilt angle : $0°$
Diameter : 61 m	Rated power : 2 000 kW
Rated wind speed : 11.2 m/s	Cut-in speed : 5 m/s
Cut-out speed : 15.8 m/s	Maximum design speed : 56 m/s
Rotational speed : 35 r.p.m.	Tip-speed ratio : $\lambda_0 = 7.8$

Synchronous generator : 2 225 kVA, 1 800 r.p.m., 4.16 kV.
Transmission ratio : k = 51
Tower height : 40 m   Hub height : 42.5 m
Orientation : Ring gear, Hydraulic drive, yaw rate : 0.25 deg
Control by computer and hydraulic pitch actuator
Rotor weight including blades : 46 t
Nacelle with the rotor :   148 t
Tower in truss pipe :   144 t
_____
Total :   292 t
Estimated production : 6 MWh at an average wind speed of 8 m/s.

The blades are constructed of a monocoque welded steel leading edge spar and an aerodynamically contoured polyurethane foam afterbody with bonded 301 stainless steel skins. The hollow one-piece steel spar which is the prime load carrier of the blade, was fabricated by welding formed ASME SA 533 steel plate. The blades are attached to the hub through a three-row roller bearing that permits the pitch angle of each blade to be varied $105°$ from feather to full power.

The gear-box and generator are similar in design to the MOD.0A, but are obviously much larger.

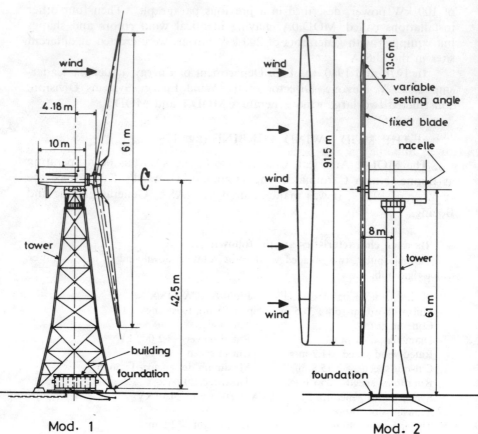

Mod. 1

Mod. 2

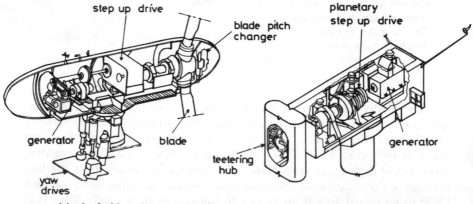

Mod. 1 Nacelle

Mod. 2 Nacelle

Fig. 179 – *Latest American wind generators.*

No significant technical or performance problems have been encountered since the starting of the wind plant.

b) THE MOD.2 WIND TURBINE (fig. 179)

The MOD.2 wind turbine is one of the largest wind turbines in the world. It operates near Goldendale, Washington, over Goodnoe Hills just north of the Columbia River Gorge.

The MOD.2 is designed to be completely computer-operated and unattended. Sensors, anemometers and other equipment on the blades, tower, and the nacelle, detect windspeed at different heights above the ground and other important details such as ice loading and potential metal fatigue.

The information from the sensor is fed into a small computer called a microprocessor located in the nacelle of the turbine. The microprocessor, using these data, automatically keeps the blades turned into the wind, starts and stops the turbine generator and changes the pitch of controllable tips of the blades to maximize power output under varying wind conditions. Should any part of the wind turbine suffer damage or malfunction, the microprocessor will immediately shut the machine down. When that happens, technicians call up information on a remote terminal, pinpointing the problem which caused the wind turbine to shut down. If the problem does not require attention from maintenance people, the microprocessor is instructed to restart the machine.

The main characteristics of the MOD.2 wind turbine are the following:

Two-bladed wind rotor upwind of the tower.
The blades are straight and made of steel. Only a part of the blade near the tip (outer 30 % of the span) can rotate or pitch to control rotational speed and power.

Diameter : 91.50 m	Power : 2 500 kW
Rated wind speed : 12.5 m/s	Cut-in speed : 4 m/s
Cut-out speed : 15.8 m/s	Maximum designed speed : 56 m/s
Rotational speed : 17.5 r.p.m.	Design tip-speed ratio : $\lambda_0 = 6.7$
Synchronous generator : 2 500 kW.	1 800 r.p.m.
Transmission ratio : k = 103	

Tower made of steel, cylindrical shape with flared base

Height : 61 m	Diameter at base : 6.4 m
Diameter from 15 to 61 m : 3.05 m	
Nacelle : Length : 11 m	Height : 2.75 m
Rotor weight :	48 t
Nacelle without the rotor :	94 t
Tower :	177.5 t
Total :	309.5 tons

The MOD.2 machine was designed using a new technology. It

incorporates a teetering rotor. This is a rigid rotor system attached to the drive shaft via a hinge axis perpendicular to the blade span. The teetering rotor reduces vibratory loads on the blades and those transmitted to the tower.

The MOD.2 tower is designed to be "soft". The softness of the tower refers to the first mode natural frequency of the tower in bending relative to the operating frequency of the system. For a two-bladed rotor, the tower is "excited" twice per rotor revolution.

For the MOD.2 the resonant frequency of the tower is between n and 2n, n being the number of rotor revolutions per second. The natural frequency of the tower is sufficiently displaced from the primary forcing frequency 2n so as not to resonate. Care has also been taken to avoid higher-mode resonance. Towers which are "soft" are less expensive than "stiff" towers but they require more accurate dynamic analysis.

The use of a new technology and design has caused a sizable decrease in the cost price of wind energy. In chapter IX, we shall see that the kWh cost price has been halved from MOD.1 to MOD.2.

13. SWEDISH WIND TURBINES (fig. 180)

The Swedish Government has decided to build two large wind turbines under the leadership of Staffan Engström, Program manager :

— the first one operates at Maglarp on Sweden's south coast near Malmö,

— the second one is located at Nässudden on the island of Gotland in the Baltic Sea.

Each of them uses a two-bladed rotor made of steel and fibre-glass reinforced plastic with variable pitch.

Their basic characteristics are given in table 27.

TABLE 27

Site	Maglarp	Gotland
Rotors	Two-bladed with variable pitch	
Position	downwind of the tower	upwind of the tower
Hub type	articulated	rigid
Diameter	78 m	75 m
Power	3 000 kW	2 500 kW
Rated wind speed	13 m/s	13 m/s
Cut-in speed	6 m/s	6 m/s
Cut-out	21 m/s	21 m/s
Rotor speed	25 r.p.m.	
Generator	alternator 3 300 kVA 1 500 r.p.m.-6.6 kV	induction generator

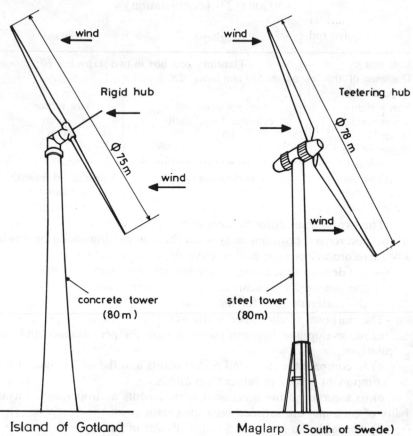

wind

wind

Rigid hub

Teetering hub

Ø 75 m

Ø 78 m

wind

wind

concrete tower
(80 m)

steel tower
(80 m)

Island of Gotland

Maglarp (South of Swede)

Fig. 180 – *Swedish wind generators.*

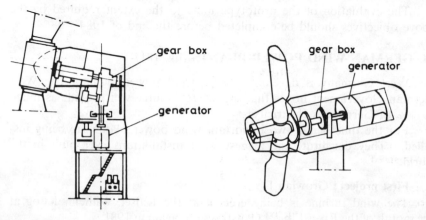

gear box

gear box

generator

generator

TABLE 27 (continuation)

Site	Maglarp	Gotland
Gear box	Planetary gear box in two stages k = 60	
Diameter of the low speed shaft	outer : 530 mm inner : 250 mm	
Yaw system	free, self-orienting	Yaw motor
Tower material	cylindrical steel shell	Concrete
Height	80 m	80 m
Weight of the nacelle	66 tons	
Contractor	Karlskronavarvet	Karlstads Mekaniska Verkstad

The two systems differ because of :

— the rotors' positions relative to the towers (downwind or upwind) and therefore by the system of orientation,

— the design of the rotors : teetering and rigid hub,

— the generators : synchronous and asynchronous,

— the materials used for the tower.

The purposes of the experiments are :

a) to develop and improve the data base for performance and stress calculations,

b) to compare the two wind power plants and the advantages of using one component or system rather than another,

c) to ascertain if the functional requirements and operational availability of each unit are satisfied on a long term basis,

d) to enable the National Swedish Board of Energy to present to the Parliament improved data for future decisions concerning eventual large-scale implementation of wind energy into the Swedish power system.

The evaluation of the prototype units to the extent required by the above objectives should be completed before the end of 1984.

14. GERMAN WIND POWER PLANTS (fig. 181)

West Germany is developing large wind power plants. The West German Government hopes that this energy source will provide as much as 8 % of Germany's electricity.

For the time being, two important wind power plants are being installed. They constitute the biggest wind installations ever built in the world.

First project : Growian I

The wind turbine is being erected at the Kaiser Wilhelm Koog at the mouth of the River Elbe. Construction began in 1981.

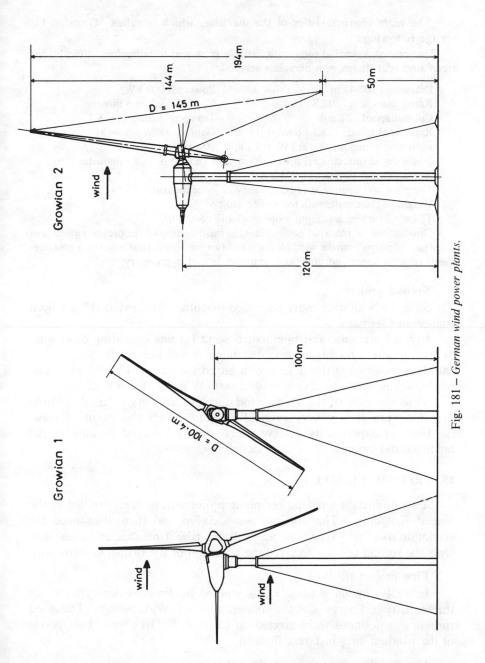

Fig. 181 – German wind power plants.

**The main characteristics of the machine, which is called "Growian I",
are tge following :**

Two-bladed downxind rotor with variable pitch and teetering hub. The blades
are of steel spar design with fibreglass aerofoil.

Diameter : 100.40 m	Power : 3 000 kW
Rated wind speed : 11.8 m/s	Cut-in speed : 6.3 m/s
Cut-out speed : 24 m/s	Tip speed ratio : $\lambda_0 = 8.3$
Rotational speed : 18.5 r.p.m. + 15 %	Gear-up ratio : k = 81

Induction generator, 3 000 kW, 6.3 kV, 1 500 r.p.m.
Single guyed tubular steel tower 100 m high and 3.5 m outer diameter.
Annual energy output : 12 GWh.
The plant will operate at rated capacity 27 % of the time.
Weight of the nacelle with the rotor : 310 t.
The wind turbine is computer-operated and controlled.
Downstream of the wind power plant, in the direction of the prevailing winds at
a distance of 500 m from the wind plants, stand two meteorological towers for measure-
ments of wind speed and direction, temperature and hygrometry.

Second project

Since 1978 another more advanced machine : " Growian II" has been
studied in Germany.

It is a horizontal-axis one-bladed wind turbine operating downwind
of its support. Its blade describes a circle of 145 m diameter. The
nacelle is mounted at the summit of a guyed steel mast, 120 m high. The
wind machine is designed to provide 5 000 kW in an 11 m/s wind.

Thus the sizes of the above wind turbines are very unusual. Moreover Growian II is a very original design. From this point of view,
the German experiments deserve to be closely watched because of the
boldness and originality of their designs.

15. BRITISH PROJECTS

Two important wind power plant projects have been studied in the
United Kingdom. The first one was delayed and then abandoned for
economic reasons. However, we shall describe it because of its features.
Only the second one has received the go-ahead of the British Government.

First project (fig. 182a) :

In 1950, a wind generator was studied by Folland Aircraft Ltd for
the Ministry of Energy with the cooperation of E.W. Golding. This wind
turbine was designed to be erected on Costa Hill, Orkney (Orkney is one
of the windiest sites in Great Britain).

Its main characteristics were the following:

Wind rotor rotating upwind of a tripod.

Two variable-pitched blades controlled by ailerons

Diameter : 68.5 m
Rated power : 3 670 kW
Rated wind speed : 15.5 m/s
Rotational speed : 42.5 r.p.m.
Generator : a.c.; induction type
Tower : Rotatable tripod : two legs carried on bogeys running on a circular rail track, the base of the third leg being at the centre of the track circle.
Hub Height : 40 m
Transmission : spur gears, direct drive.
Start and stop by pilot windmill, yaw by fantail coupled to bogey wheels through centrifugal clutch.

Second project (fig. 182b) :

The studies began in 1976. They were undertaken by a group comprising Taylor Woodrow Construction Limited, British Aerospace, GEC and ERA. The decision to erect the 3 MW wind generator on Burgar Hill, Orkney Island, was taken by the British Ministry of Energy in January 1981.

The annual mean wind speed on Burgar Hill at the hub height of the 3 MW wind turbine has been estimated to be more than 12 m/s.

The main characteristics of the wind turbine are the following :

Two-bladed wind rotor made of steel operating upwind of a circular tower. Fixed-pitch blades except for the outer 20 % of the span.

Diameter : 60 m Cut-in speed : 7 m/s
Power : 3 000 kW Cut-out : 27 m/s
Rated wind speed : 17 m/s Synchronous generator :
Rotational speed : 30 r.p.m. Annual energy output : 10.5 GWh

The rotor, transmission and generator will be mounted at the top of a steel cylinder, 3.5 m in diameter and one inch thick with a natural frequency 1.4 times the rotational frequency constituting therefore a "soft" tower. The tectered rotor rotating at a height of 45 m above ground level will drive a synchronous generator via a gearbox. This generator will be placed vertically thus minimising the volume of the nacelle and eliminating the need for power slip rings.

Regulation of power output will be made through the springs and dampers supporting the gearbox and the partial span variable pitch blades hydraulically controlled.

In case of very strong winds, braking of the machine will be obtained by variation of the tip-blade pitch and feathering. At 10 % full rotational speed, the mechanical brake will be applied to stop the machine.

Performance measurements and monitoring of the system will be accomplished through a microprocessor.

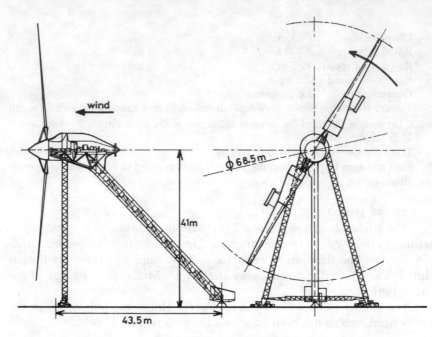

Fig. 182a – *First British design (1950)*
(According to Golding, Spon LTD publishers)

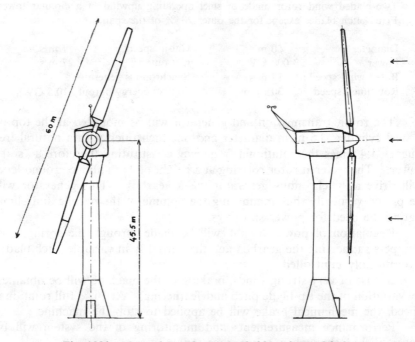

Fig. 182b – *Second British project (liable to be modified).*

The machine is designed for survival gust speeds of 70 m/s when stationary and 40 m/s when operating.

Initially, the wind power plant was designed with an induction generator. But as the electrical grid of Orkney is " weak ", some changes had to be made in the design, and a synchronous generator was chosen.

Machine of 20 m diameter

The 3 MW machine will share the same site as a smaller 20 m diameter version which will have a rated output of 250 kW. This is expected to be put into operation in winter 1981/1982 serving as a development prototype for the larger machine. It will be linked to a microprocessor for control. The 20 m diameter machine is similar to the larger machine except for the tower which is "stiff "(high frequency).

The power train drives a synchronous generator which, for the purpose of research and development, can be operated in two alternative ways :

— Directly connected to the grid with the rotor running at constant synchronous speed. Power regulation is achieved by soft drive and by operation of the moving blade tips which take up 20 % of the length of each blade.

— Connected to the grid through a power conditioning unit so that the rotor can run at varying speed whilst feeding power into the grid at constant frequency. Power peaks from gusts are absorbed by allowing the rotor to accelerate.

The 20 m diameter machine is designed with a teetered hab that may be lockes, so that experience of both teetered and rigid hab machines may be gaines.

16. CANADIAN PROJECT

The Hydro-Quebec Company and the National Research council of Canada intend to build a very powerful two-bladed Darrieus rotor in the shape of an egg-beater like the wind generator on the Madeleine Islands. This wind generator will have a total height of 110 m, a useful height of 96 m, a maximal diameter of 64 m and a capacity of 3,8 megawatts. If all goes well, the wind turbine will be on line by the end of 1983.

17. ANOTHER DESIGN

In 1933, the German Honneff suggested the construction of a metallic lattice tower, 300 m high. He designed it to support five wind turbines of 75 m diameter, able to provide together 50,000 kW of power. Honneff had been impressed by the increase of wind velocity above the level ground.

18. TECHNICAL CONSIDERATIONS AND CONCLUSIONS

The present chapter shows that wind power installation technology has greatly improved in the last few years.

What conclusions can be drawn from the various experiments which have been carried out around the world?

For the wind power plant whose diameter is less than 30 m or 40 m, a rigid hub can be adopted, especially if the rotor is three-bladed, because this type of rotor is better balanced than the two-bladed one. The pitching moment for the former is smaller. Consequently, the bending stresses inside the rotor shaft are reduced.

If the diameter is greater than 40 m, it seems that it is better to choose a teetering hub with a two-bladed rotor. Thus the problem of the pitching moment is eliminated. The stresses inside the blades and the load on the tower are reduced considerably.

Should the rotor be placed upwind or downwind of the support?

Both possibilities may be argued.

If the tower is cylindrical with a large diameter, the rotor should be located upwind of the support because of the extent of the tower shadow in the downwind position. With a lattice tower or a cylinder of small diameter, it is possible to place the rotor downwind.

Concerning the rotor, the regulation and the safety of the installation, which is the best system? Is it preferable to choose fixed blades with spoilers, or variable-pitch blades feathering in strong winds?

The first system has the advantage of strength. The second is more fragile but control is total.

The Danish and American experiments show that the best solution consists of a rotor with blades partially controlled in pitch. According to Helge Petersen, the inner part of the blades must be fixed and made of steel and GRP, the outer part (30 % to 50 % of the full span) being movable and controlled in pitch and made of fibreglass-reinforced plastic (GRP) in order to reduce the weight.

Why choose steel for constructing the inner part of the blades? Steel is a more reliable material than GRP. Moreover as steel is more rigid than GRP, in normal running, the displacement of the blade tip backwards under aerodynamic forces is reduced by about half if steel is used. Connecting steel blades to the hub and together is also easier.

The regulation will be a feathering regulation. In case of overspeeding, this means that the chords of the outer part of the blades will take a position parallel to the wind velocity.

According to Helge Petersen, stalling regulation gives rise to vibrations because of oscillating and eddying wakes created on the extrados of the blades.

Of course, every wind power plant must be provided with a mechanical brake for stopping the rotor.

The orienting system: If the rotor is located upwind of the tower, a yawing motor is necessary. If the rotor is placed downwind of the support, the machine can be self-orienting. In this case, to avoid oscillations about the vertical axis, it is advisable to install a damping system. However, a yawing motor can also be used. This arrangement is better if the distance from the pivot to the rotation plane is very short.

The support: The most economical model is the low-frequency type. It may consist either of a steel tubular tower or a concrete tower. It is lighter but not so strong as the high-frequency model. This explains the fact that high-frequency models are always constructed. But whatever the type chosen, it should be noted that the support must withstand the loss of a blade. Thus it has to be designed for this eventuality. This is an important point for the safety of the wind power plant. As a matter of fact, statistics concerning wind energy systems show that many wind power installations have been stopped by the breaking of a blade. Thus, if the preceding precautions have been observed, the installation will not be completely destroyed if a blade breaks. In Sweden, all high power wind installations are designed to withstand the centrifugal force which would result at the head of the mast if a blade broke at its root.

CHAPTER VIII

WIND POWER PLANT PROJECTS

In this section, we apply the theories presented in the former chapters to the designs of four different wind power installations :
— a water pumping station using a Savonius rotor,
— a low-speed wind turbine coupled to a piston pump,
— a horizontal-axis aerogenerator,
— a Darrieus rotor driving an electrical generator.

We have limited each design to the aerodynamic calculations of the blades, to the determination of the gear-box ratio and to the evaluation of the performance. To avoid lengthy developments, the strength of materials and the regulation problems have been deliberately left out.

In addition, we have assumed that the statistical study of the local winds has been performed, making it possible to state accurately the wind turbine model to be used and the rated wind velocity to be considered for determining the geometrical characteristics of the wind rotor. The maximal efficiency of a well-matched rotor must occur at about the same wind speed as the peak in the wind energy distribution.

When no statistical study is available, the rated wind speed will be fixed at about 1.5 times the average annual wind velocity.

1. DESIGN OF A WATER PUMPING STATION USING A SAVONIUS ROTOR

We intend to build a water pumping station using a Savonius rotor. Two drums of 230 l capacity each, commonly found in gas stations have been used to manufacture the Savonius rotor. Each drum has been cut diametrically into two parts and the parts then welded together. The measurements of the original drums were as follows :
Height : h = 0.90 m. Diameter : d = 0.58 m.

Required conditions

The rotor must start at a wind velocity equal to 2.50 m/s. It drives a piston pump which lifts the water to a height of 10 m. The power efficiency of the whole set is about 70 %. The water discharge for 3, 4, 5 and 6 m/s wind velocities will be stated.

Solution

Design of the rotor

We shall adopt (see fig. 99) the value : $e = \dfrac{d}{6} = 10$ cm.

For this value, the performance of the Savonius rotor is optimal.

Thus for the four half-drums, the maximum area which is intercepted on the windstream, is equal to :

$$S = 2\,hD = 2\,h\,(2d - e) = 1.9 \text{ m}^2$$

To facilitate starting, the rotor will be constructed in two stages welded at right angle.

To determine the performance, we shall use the results obtained for the rotor II (fig. 9) by the Canadians and especially the curve of variation of the torque coefficient C_m as a function of the speed ratio λ_0.

Given :

ρ : the specific mass of the air $\rho = 1.25$ kg/m³,
V : the wind speed in m/s,
N : the rotational speed of the rotor in revolutions per second,
k : the gear-up ratio ($k = n_1/n_2$, see chapter VI),
One revolution of the rotor corresponds to k piston strokes.
$\bar{\omega}$: the specific weight of the water ($\bar{\omega} = 9\,800$ N/m³),
q : the volume of water extracted at every piston stroke,
η : the power efficiency.

The power absorbed by water pumping is equal to :

$$P_a = \frac{\bar{\omega}kNqH}{\eta}$$

Thus the corresponding torque is :

$$C_R = \frac{\bar{\omega}kNqH}{\eta\,2\pi N} = \frac{\bar{\omega}}{2\pi\eta}\,kqH$$

When the rotor is running normally, the aerodynamic torque M is equal to the pumping torque C_R :

$$M = \frac{1}{2}\,\rho C_m RSV^2 = C_R$$

If η is constant and k fixed, C_R keeps a constant value.

To obtain the values of the discharge as a function of the wind speed, we can apply the method indicated in Chapter VI. The curves of variation

of the torque M as a function of the rotational speed for different wind velocities may be drawn from the characteristic $C_m(\lambda_0)$ shown in fig. 99.   The
intersections of these curves with the horizontal line whose ordinate is C_R,
determine the rotational speed N for the different wind velocities.   As Q
and N are related by the equation $Q = kNq$, it is possible to obtain the discharge Q for each value of V and therefore, the law of variation $Q(V)$.

In practice, the calculations may be shortened.

Instead of plotting the former curves, we can express the equality
between M and C_R as follows:

$$\frac{\bar{\omega}}{2\pi\eta} \, kqH = \frac{1}{2} \, \rho C_m RSV^2$$

The preceeding relationship may be written as:

$$C_m = \frac{\bar{\omega}}{2\pi\eta} \, \frac{kqH}{RSV^2}$$

Let Y be the quantity:

$$Y = \frac{\bar{\omega}}{2\pi\eta} \, \frac{kqH}{RSV^2}$$

Replacing $\bar{\omega}$, η, ρ by their values, we obtain:

$$Y = 3\,500 \, \frac{kqH}{V^2}$$

Thus the problem is reduced to the determination of the points of
intersection of the curve $C_m(\lambda_0)$ with horizontals whose ordinates are
proportional to $1/V^2$.   Solving this problem is not difficult (see fig. 184·)

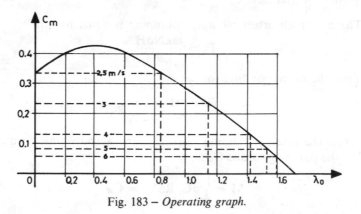

Fig. 183 – *Operating graph.*

It must be pointed out that the quantity kqH is determined by the
starting conditions.   The rotor will run when the aerodynamic starting

torque is slightly higher than the pumping torque. We shall assume that the value of the latter is equal to its average value.

The coefficient C_m for the starting torque is equal to 0.35 ($\lambda_0 = 0$). Therefore, the quantity kqH will have to meet the equation:

$$0.35 = 3\ 500\ \frac{kqH}{V^2}$$

for a wind velocity of 2.5 m/s which is required at starting point.

The former expression may be reduced to:

$$kqH = 0.62\ 10^{-3}$$

As $H = 10$ m, we get for the quantity kq:

$$kq = 0.62\ 10^{-4}\ m^3$$

Now the rotor is running. If the wind velocity keeps a constant value of 2.5 m/s, the rotational speed of the rotor will stabilize at a value such that λ_0 will be equal to 0.85. (This value corresponds to the intersection of the curve $C_m(\lambda_0)$ with the horizontal line whose ordinate Y relates to the starting wind velocity $V = 2.50$ m/s.

The rotational speed N and the discharge Q obtained will respectively be equal to:

$$N = \frac{\lambda_0 V}{2\pi R} = \frac{0.84 \times 2.50}{6.28 \times 0.53} = 0.63\ \text{r.p.s.} = 38\ \text{r.p.m.}$$

$$Q = kNq = 0.62.\ 10^{-4} \times 0.63 = 0.39\ 10^{-4}\ m/s = 0.039\ l/s$$
$$= 140\ l/h.$$

Now suppose the wind speed increases and becomes 3 m/s. Y takes the value:

$$Y = \frac{3\ 500 \times 0.62.\ 10^{-3}}{9} = 0.24$$

The speed ratio relative to the intersection of the corresponding horizontal line with the curve $C_m(\lambda_0)$ reaches the value $\lambda_0 = 1.16$. This corresponds to a rotational speed N such that:

$$N = \frac{1.26 \times 3}{6.28 \times 0.53} = 1.05\ \text{r.p.s.} = 63\ \text{r.p.m.}$$

and to a discharge:

$$Q = 0.62.\ 10^{-4} \times 1.05 = 0.65.\ 10^{-4}\ m^3/s = 0.065\ l/s = 214\ l/h$$

The results obtained at different wind velocities and the corresponding power coefficients are given in table 28.

TABLE 28

V m/s	$C_m = Y$	λ_0	r.p.m.	Q l/h	C_p
2.50	0.35	0.84	38	140	0.294
3	0.24	1.16	63	234	0.278
4	0.135	1.42	105	382	0.192
5	0.087	1.54	140	518	0.134
6	0.060	1.60	174	650	0.096

When the wind velocity increases, the power coefficient diminishes.

Moreover table 28 shows that the rotational speed of the rotor can reach 174 r.p.m. at a wind velocity equal to 6 m/s. But good maintenance of the water piston pump requires that the number of strokes never exceed 40 or 50 per minute. Therefore, we shall adopt for the gear-box ratio: k = 1/4. This will limit the piston speed to 43.5 strokes per minute at a wind velocity of 6 m/s.

From the relationship: $kq = 0.62 \cdot 10^{-4}$ m³, we obtain: $q = 4 \times 62 = 248$ cm³ for the cylinder capacity of the pump.

Assuming the internal diameter of the pump is 5 cm (corresponding section: 19.6 cm²), the stroke of the piston will be equal to:

$$248/19.6 = 12.7 \text{ cm.}$$

Therefore, the length of the crank arm will be chosen equal to:

12.7: 2 = 6.35 cm.

Remarks

1º For values of lift H other than 10 m, it is not necessary to repeat all the calculations. The new discharges and the new values for the cylinder capacity are obtained by multiplying the discharges and the cylinder capacities q of the pump given in table 26 by the ratio 10/H. The rotational speeds remain unaltered.

2º If the rotor is coupled to a membrane pump, the calculation is done in a similar manner. The only difference is that a membrane pump may be subjected to higher speeds than a piston pump (up to 200 strokes per minute) so reduction gearing will not be necessary.

3º In the above section, we have not dealt with the regularization of the pumping torque. For the rotor to start easily it is indispensable that this regularization be assured. A spring or a counterweight will be used to get the balancing effect.

2. MULTIBLADED WIND TURBINE DRIVING A PISTON PUMP

Let us determine the main geometrical characteristics of a multibladed wind machine driving a piston pump according to the following conditions:

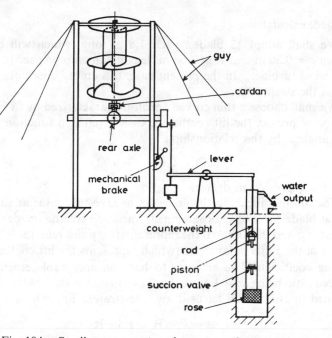

Fig. 184 – *Small water pumping plant using a Savonius rotor.*

Hydraulic power to be delivered at 5 m/s wind speed : 330 W
Mechanical efficiency : η = 65 %
Optimal tip-speed ratio : λ_0 = 1.5

Determination of the diameter and the rotational speed

As the hydraulic power and the mechanical efficiency are respectively 330 W and 65 %, the mechanical power of the wind turbine is equal to :

$$P = \frac{330}{0.65} = 508 \text{ W}$$

Calculate the diameter by applying the relationship : $P = 0.15 \, D^2 V^3$. Replacing P by its value for V = 5 m/s, we get : D = 5.2 m.
Assuming the wind rotor is operating with a tip-speed ratio λ_0 = 1.5 at V = 5 m/s, we obtain for the value of the rotational speed :

$$N = \frac{1.5 \times 5}{\pi \times 5.2} = 0.46 \text{ r.p.s.} = 27.5 \text{ r.p.m.}$$

It is advisable that a piston pump not exceed 40 or 50 strokes per minute. Taking this into account as well as the former rotational speed and the fact that the wind turbine will be equipped with a regulating system coming into operation at about 6.5 m/s, it is not necessary to fit the wind rotor shaft with a reduction gear to drive the pump. The latter may be driven directly by the wind turbine.

Blade calculations

We shall adopt 12 blades, each 1.8 m long, which will be fastened between r = 0.80 m and R = 2.60 m, leaving an empty space in the middle of the wind turbine.   In the present case, this empty area does not exceed 10 % of the swept area.

We shall choose a thin curved aerofoil characterized by f/l = 0.1.   For this type of profile, the lift coefficient C_l is given as a function of the incidence angle i, by the relationship :

$$C_l = 0.074\, i + 0.62$$

where i is expressed in degrees.

The optimal incidence being 3°, let us take this value as the incidence angle at blade tip, and choose a linear variation for the incidence between the blade tip and the hub.   If the representative point remains in the increasing part of the Eiffel polar curve (which represents the lift coefficient versus the drag coefficient), we are sure to have an acceptable efficiency.   The tip-speed ratio being $\lambda_0 = 1.5$, on the axes situated at the base of the diagram contained in chapter IV let us draw, the straight line whose equation is :

$$\lambda = \lambda_0\, r/R = 1.5\, r/R$$

Following the arrows drawn on the diagram, we obtain for each section the angle I and the quantity $C_l bl/r$, then the setting angle, by substracting the incidence angle i from the angle I.

The values obtained for the different parameters are collected in

TABLE 29

$\dfrac{r}{R}$	λ	$I°$	$i°$	$\alpha° = I - i$	C_l	$\dfrac{C_l bl}{r}$	l in m
0.3	0.45	44	6.5	37.5	1.12	7.1	0.42
0.4	0.6	39.4	6	33.4	1.08	5.7	0.46
0.6	0.9	32	5	27	1.02	3.82	0.485
0.8	1.2	26.5	4	22.5	0.94	2.65	0.485
1.0	1.5	22.4	3	19.4	0.97	1.72	0.43

In practice, the blades will be fastened on curved iron supports at distances from the hub axis equal to r = 0.4 R and r = 0.8 R.   The curved iron supports in those places will be fixed in such a manner that the setting angles are respectively 33.4° and 22.5°, the chords of the blade in those sections being 0.46 m and 0.49 m and the depths of curvature 4.6 cm and 4.9 cm.

Calculation and design of the tail vane

For the distance L between the support and the middle of the tail vane, we shall adopt a value approximately equal to 60 % of the diameter L = 3.20 m, and for the distance E between the support and the plane of rotation of the rotor: E = L/4 = 0.75 m.

The area A of the tail vane will be chosen equal to 10 % of the swept area: A = 2.15 m².

The regulation will be assured by means of a joint of the connecting rod to the tail vane, a spring regulator coupled to a damper, and by shifting laterally the axis of rotation of the wind machine relative to its pivot by 10 cm.

Design of the pump for lifting water to a height of 10 m

For a wind speed of 5 m/s, the hydraulic power is 330 W and the rotational speed of the windmill, 27.5 r.p.m.

Let q be the cylinder capacity of the pump. The discharge of water Q extracted per second is given by the equation:

$$Q = \frac{27.5}{60} q$$

and the hydraulic power delivered by:

$$P_H = \bar{\omega}QH = 330 \text{ W}$$

From the previous equations, we get:

$$q = \frac{60 \times 330}{27.5 \times 9\,800 \times 10} = 0.0073 \text{ m}^3 = 7.3 \text{ dm}^3$$

We shall adopt a single-acting pump. The smoothing of the torque will be assured by a strong spring.

Taking a value of 30 cm for the pump stroke, we obtain for the piston section S = 7.3/3 = 2.44 dm² which corresponds to a bore equal to 17.6 cm.

For a lift other than 10 m, the cylinder capacity will be taken as 73/H. The discharge for equal wind velocities will be then reduced proportionally to 10/H.

If a reduction gear having a gear down ratio equal to two was inserted between the wind machine and the pump, it would be necessary to double either the stroke or the piston section. In such a case, the average discharge would be kept the same.

At a wind velocity of 7 m/s, the windmill can develop a maximal power of 1 400 Watts.

3. CALCULATION AND DESIGN OF A HORIZONTAL-AXIS WIND TURBINE DRIVING AN ALTERNATOR

Determine the main measurements of a three-bladed aerogenerator required to provide an electric power of 8 kW at a wind velocity of 8 m/s (rated wind speed) and having a maximum efficiency for a tip-speed ratio $\lambda_0 = 7$.

Rated speed of rotation of the alternator: 1 500 r.p.m.

Solution

Diameter

Assuming that the efficiency of the alternator and the step-up gearing is 80 %, the mechanical power provided on the rotor shaft at 10 m/s is equal to:

$$P = 8 : 0.8 = 10 \text{ kW}$$

By applying the relationship: $P = 0.2 \, D^2 V^3$ which holds for a high-speed wind machine, we obtain for the diameter D:

$$D = \frac{P}{0.2 \, V^3} = \frac{1\ 000}{0.2 \times 512} = 9.88 \text{ m}$$

We shall adopt: $D = 10$ m.

Rotational speed of the wind rotor

The aerodynamical efficiency of the wind rotor must be optimal for $\lambda_0 = 7$. For a wind speed of 8 m/s, this corresponds to a rotational speed of the wind rotor equal to:

$$N = \frac{\lambda_0 V}{\pi D} = \frac{7 \times 8}{\pi \times 10} = 1.78 \text{ r.p.s.} = 107 \text{ r.p.m.}$$

Gear-up ratio

The rated speed of rotation of the alternator is 1 500 r.p.m., So it is necessary to insert a step-up gearing between the wind rotor and the alternator. The gear-up ratio k will be chosen equal to:

$$k = \frac{1\ 500}{107} = 14$$

Calculation and design of the blades

Profile : We shall adopt for the profiles of the section the NACA 23015 aerofoil. The variations of the lift and drag coefficients as a function of the incidence angle are given by the following relationships:

$$C_l(i) = 0.1 + 0.11 \, i \text{ for } i < 10°$$

$$C_l(i) = 1.5 - 0.0188 \, (i - 14)^2 \text{ for } 10° < i < 15°$$
$$C_d(i) = 0.007 + 0.0055 \, (C_l(i) - 0.2)^2 \text{ for } i < 10°$$
$$C_d(i) = 0.0125 + 0.16 \, (C_l(i) - 1.1)^2 \text{ for } i > 10°$$

The most favourable angle of incidence (maximum efficiency) is about 6°.

For the law of variation of the incidence angle as a function of the radius, we shall choose the expression :

$$i = 12.5 - 7.5 \, \frac{r}{R}.$$

The value of the angle of incidence varies from 5.25° at the blade tip to 12° at 0.1 R distance from the rotation axis. These values of incidence angles are related to the increasing part of the curve C_l/C_d. We can thus be confident that the wind machine will have a high aerodynamic efficiency.

The determination of the inclination angles and the quantities $C_l bl/r$ has been carried out with table 7 according to the method laid out in chapter IV.

The inclination and incidence angles being known along the blade, the setting angles may very easily be obtained ($\alpha = I - i$) as well as the chords of the blades for the different sections. The results are set out in table 30.

TABLE 30

$\dfrac{r}{R}$	λ	$I°$	$i°$	$\alpha°$	C_l	$\dfrac{C_l bl}{r}$	l in m
0.1	0.7	36.67	12	24.67	1.425	4.975	0.582
0.2	1.4	23.69	11.25	12.44	1.358	2.12	0.52
0.3	2.1	16.98	10.5	6.48	1.270	1.095	0.431
0.4	2.8	13.10	9.75	3.35	1.172	0.654	0.372
0.5	3.5	10.63	9.00	1.63	1.090	0.431	0.330
0.6	4.2	8.93	8.25	0.68	1.007	0.305	0.302
0.7	4.9	7.69	7.50	0.19	0.925	0.226	0.285
0.8	5.6	6.75	6.75	0	0.842	0.174	0.276
0.9	6.3	6.01	6.00	0.013	0.760	0.138	0.273
1	7	5.42	5.25	0.17	0.677	0.112	0.276

Table 30 shows that the chord diminishes progressively from the axis towards the blade tip. It may also be noted that the pitch angle α reaches a minimum value for $r = 0.8$ R and then increases. To avoid this "anomaly" and make it easier to build the blades, we shall take $\alpha = 0$ between $r = 0.8$ R and the blade tip. This compels us to replace the two last lines of the foregoing table by the following ones :

$\dfrac{r}{R}$	λ	$I°$	$i°$	$\alpha°$	C_l	$\dfrac{C_l bl}{r}$	l in m
0.9	6.3	6.01	6.01	0	0.761	0.138	0.272
1	7	5.42	5.42	0	0.696	0.112	0.268

The corrections are very slight.

Performance of the wind machine

The geometrical characteristics and the performance of the wind rotor have been determined with a computer according to the AERO program given in the appendix.

The program has enabled us to get the characteristics $C_m(\lambda_0)$ and $C_p(\lambda_0)$. The maximum value of C_p (C_p max = 0.485) has been obtained for $\lambda_0 = 7$. This value allows us to predict good results.

Design of blades having straight leading and trailing edges

The making of the above blades does not pose any difficulty if the blades are made of plastic, because this material is easily shaped. But it is not so easy if steel or aluminium are used. In such cases, it is convenient to choose blades with a straight leading edge and a straight trailing edge. To find the new shape of the blades, we have started from the previous results and have assumed that the sections situated at distances 0.2 R, 0.5 R and 0.9 R from the rotation axis remain unaltered as to position and measurements.

Thus the tip and intermediate profiles have been obtained by alignment of their leading edge and trailing edge on the unaltered sections.

This operation and the new characteristics $C_m(\lambda_0)$ and $C_p(\lambda_0)$ have been computed by the EOLE program given in the appendix.

In that program, I_1, I_2, I_3 represent the numbers characterizing the position of the basic sections. Here I_1, I_2, I_3 are respectively equal to 2, 5, and 9.

The results are collected in Table 31.

TABLE 31

$\dfrac{r}{R}$	λ	$I°$	$i°$	$\alpha°$	l_1 in m
0.1	0.7	36.67	12	24.67	0.586
0.2	1.4	23.69	11.25	12.44	0.52
0.3	2.1	16.98	10.5	6.48	0.455

0.4	2.8	13.10	9.75	3.35	0.391
0.5	3.5	10.63	9.00	1.63	0.330
0.6	4.2	8.93	7.64	1.28	0.316
0.7	4.9	7.69	6.79	0.90	0.301
0.8	5.6	6.75	6.29	0.48	0.287
0.9	6.3	6.01	6.00	0.013	0.273
1	7	5.42	5.92	−0.50	0.259

The power coefficient is maximal for $\lambda_0 = 7$.
Its value has been slightly altered by the alignment:

$$C_p \ \text{max} = 0.483$$

It must be pointed out that the EOLE program enables us to choose only two basic sections instead of three.

Layout

We shall choose:
— a variable-pitch wind generator if the wind installation is planned to run autonomously. The rotational speed will be controlled by a regulator like the Aerowatt model. This type of regulator enables the wind machine to start easily,
— a fixed-blade aerogenerator driving an induction generator if the wind plant is connected to the grid. Spoilers will be placed at the blade tips to protect the set against overspeeding. The induction generator will be used as a motor for starting when the wind speed is not high enough.

In both cases, the rotors will be equipped with automatically-actuated mechanical brakes and placed either downwind or upwind of the support. To reduce the bending stresses, the blades will be fixed on the hub, slightly inclined with respect to the plane perpendicular to the rotor shaft.

Continuation of the design

The wind-tunnel tests have given the result shown in fig. 185. Assu-

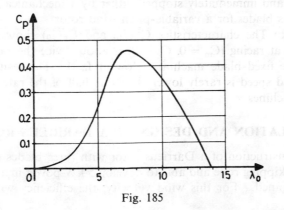

Fig. 185

ming that the generator is connected to the grid, calculate the power provided by the wind plant at wind velocities of 4, 6, 8, 10, 12, 14, 16, 20 m/s.

If the generator is an alternator, the wind rotor is rotating at 107 r.p.m. and slightly more if it is an induction generator. In practice, for the latter case, we can considerer that the rotational speed of the wind rotor is 107. r.p.m. Thus the speed of the blade-tips is :

$$U_0 = \pi ND = \pi \times 10 \times 107/60 = 56 \text{ m/s}$$

We can evaluate the tip-speed ratios $\lambda_0 = U_0/V$ at the different wind velocities, and determine the corresponding power coefficients from fig. 185, then the mechanical and electrical powers supplied by the wind generator by the relationships :

$$P = \frac{1}{2} \rho C_p S V^3 \quad \text{and} \quad P_e = 0.8 \text{ P.}$$

The results appear in table 32.

TABLE 32

V, m/s	4	6	8	10	12	14	16	20
λ_0	14	9.33	7	5.6	4.66	4	3.5	2.8
C_p	0	0.39	0.46	0.32	0.20	0.13	0.09	0.05
P, kW	0	4.13	11.5	15.7	17.0	17.5	19.1	19.6
P_e, kW	0	3.30	9.20	12.6	13.6	14.0	14.5	15.7

Therefore we shall choose an induction generator of 15 kW capacity, 1 500 r.p.m. It will be automatically disconnected from the grid at 16 m/s wind speed and immediately stopped either by a mechanical brake, or by feathering its blades for a variable-pitch wind rotor.

Remark : The characteristics $C_m(\lambda_0)$ and $C_p(\lambda_0)$ show that the tip-speed ratio at racing ($C_m = 0$, $C_p = 0$) is about twice the rated tip-speed ratio for the fixed-blade machines (slow or fast). The result is that the starting wind speed is rarely lower than one half of the rated wind speed for these machines.

4. CALCULATION AND DESIGN OF A DARRIEUS ROTOR

The construction of a Darrieus rotor with three blades curved in the shape of a skipping rope and able to develop 5 kW power in a 7.5 m/s wind speed is planned. For this wind velocity, the efficiency will have to be optimal and the rotational speed of the vertical-axis wind turbine equal to 85 r.p.m.   The profile of the blades consists of a NACA 0012 aerofoil and the maximum radius of the turbine is equal to half the rotor height.

Determine the measurements of the wind rotor (height, diameter of the rotor and chord of the blade).

Consider also the case of a rotor having only two blades.

Solution

The maximal power which may be supplied by a Darrieus rotor, is given approximately by the following expression:

$$P = 0.25 \ SV^3$$

P, S and V being respectively expressed in W, m^2 and m/s.
Thus the intercepted area will be:

$$S = \frac{P}{0.25 \ V^3} = \frac{5 \ 000}{0.25 \times 120} = 47.5 \ m^2.$$

The intercepted area S is given as a function of the geometrical characteristics by the expression: $S = 8 \ RH/3$,

R and H being the maximum radius and the half of the rotor height.

If R = H then $S = \frac{8}{3} R^2$ and $R = \frac{3}{8} \times 47.5 = 4.20$ m

The total height of the rotor therefore will be 8.4 m.
The operating conditions correspond to a speed ratio:

$$\lambda_0 = \frac{2\pi NR}{V} = \frac{2\pi.85.4.2}{60 \times 7.5} = 5$$

The results obtained by Jack Templin, Ottawa, Canada, have shown that the speed ratio λ_0 is approximately related to the ratio R/bl by the expression:

$$\lambda_0 = \sqrt{\frac{5R}{bl}}$$

This relation gives:

$$l = \frac{5R}{b\lambda_0^2} = \frac{5 \times 4.2}{3 \times 25} = 0.28 \ m$$

l being the chord of the blade and b, the number of blades.

As the relative thickness of the NACA aerofoil is equal to 12 %, we obtain for the maximum thickness of the blade:

$$28 \times \frac{12}{100} = 3.36 \ cm$$

For a two-bladed rotor, the chord and the thickness of the blades would be respectively $1 = 42$ cm and 5.05 cm.

To give more information about the operating conditions, it is necessary to know the generator characteristics.

The variations of the power coefficient as a function of the speed ratio may be determined by the FANN program given in the appendix or taken from the studies of Jack Templin concerning the parabolic rotor having a solidity ratio equal to :

$$\frac{bl}{R} = \frac{3 \times 0.2}{4.2} = 0.2$$

To use the program, it is necessary to know the drag coefficient CDO of the rotor rotating in a still atmosphere.

The FANN program has been established with a value : $CDO = 0.012$.

Layout

If the wind plant must be connected to the grid, we shall choose a three-phased induction generator whose capacity will be calculated in the same way as in the former project.   In the present case, a capacity of 10 to 12 kW is enough.   If the rated rotational speed of the generator is about 750 r.p.m., we shall adopt a gear-box having a ratio approaching 8.8 or 8.9. At starting, the generator will run as a motor.

If the wind plant is used autonomously, it is yet possible to use an induction generator shunted by capacitors.   The starting will be facilitated by connecting it temporarily to the grid through resistors or by using an auxiliary source of energy.   The voltage will be regulated through variable resistors electronically controlled.

In both cases, safety devices are indispensable.

CHAPTER IX

WIND POWER PLANTS:
ECONOMIC AND
DEVELOPMENT PROSPECTS

The aim of the present chapter is to examine the position of wind power plants in the world energy market in comparison with other energy sources.   Therefore, we shall here consider the cost of energy provided by different means.

1. WIND ENERGY

Wind is free but, unfortunately, wind machines are not.   Moreover from time to time, wind machines break down.   Thus they have to be repaired, which may cost a considerable amount of money.   Likewise, they incur costs of operation and routine maintenance.

Table 33 gives the cost of electricity (COE) produced by wind turbines of various diameters manufactured by different firms.

The calculations have been made assuming that the installation cost will be spread over 20 years with an annual interest rate of 15 %.   They correspond to an annual fixed charge of 20 % (16 % on capital + 4 % for annual operation and maintenance costs) with a 20 year lifetime.

Note that taking as a basis an annual interest rate of 15.8 %, a lifetime of 30 years and 4 % for operating and maintenance costs, we obtain the same annual charges.

The kWh costs have been computed by the relationship:

$$\text{COE (cents/kWh)} = \frac{20 \times (\text{Price of the wind plant, \$})}{(\text{Annual energy, kWh})}$$

For the American high-power wind installations, second-unit costs are quoted so as not to include the non-recurring costs associated with the first prototype unit.

TABLE 33 : **Small wind machines :** Assumed lifetime : 20 years

Company	Diameter in m	Power in kW	Annual output for 6 m/s mean wind speed in kWh	Price in $	Investment $/kW	kWh price cents/kWh
Aerowatt	5 9.20	1.1 4.1	5 000 20 000	20 000 38 000	18 182 9 270	80 38
Enag	2.65 5 6	1 4 6	2 000 7 000 10 000	4 550 14 545 20 000	4 550 3 636 3 484	45.5 41.5 40
Dunlite	4.10	2	3 000	5 100	2 550	34
Winco	1.8	0.2	300	645	3 225	43
Sencenbaugh	3.6	1	2 500	2 950	2 950	23.6
Northwind	5	2	5 000	7 200	3 600	29
Pinson (v-a)	5	5	5 000	7 000	1 400	28
Kuriant	10.9	15	28 000	16 500	1 060	11.8
Windmatic	12	30	45 000	29 000	970	13
Sonebjerg	12	45	45 000	24 000	550	10.6
U. Poulsen	13	5.5/30	40 000	19 000	633	9.5
Nordtank	11	7.5/22	30 000	25 100	1 140	16.8
Vesta	15	5/45	60 000	38 000	844	12.6

High-power wind machines : Assumed lifetime : 20 years

	Diameter in m	Power in kW	Annual output in MWh	Price in $ millions	Investment in $/kW	kWh price cents/kWh
Volund	29	260	500	0.36	1 380	14.4
Nibe	41	630	1 500	1.1	1 750	14.6
MOD.0A	38	200	1 200	1.61 (1977)	8 000	26.8
MOD.1	61	2 000	5 000	5.40 (1977)	2 700	21.6
MOD.2	91.5	2 500	10 000	4.3 (1980)	1 350	8.6
Marglap	78	3 000	8 000	10.5 (1978)	3 500	26.2
Gotland	75	2 000	6 000	8.7	4 250	27.4
Orkney	60	3 000	10 500	11.3	3 750	21.6
Growian	100,4	3 000	12 000	18	6 000	30

For small wind machines, table 33 shows that the cost price of the kWh varies much with the firm.   This is due to the unequal level of wind machine industrialization in the different countries.   In Denmark, where the Kuriant, Windmatic, Sonebjerg, Poulsen, Nordtank and Vesta Companies are found, the industrial construction of wind machines is very developed. During the last two years more than six hundred windmills of 10-30 kW capacity have been erected in the country.   Thus the manufacturers can sell the machines at cheaper prices because they are made in series.   In the U.S.A. and in France, the small machines are only made one by one. Therefore, they are too expensive and very few people buy them.

In the range of high-power wind machines, the cost price is minimal for the MOD.2 American wind turbine.   This is due to the important program launched by the US Department of Energy in the wind energy field.

Note that the decrease in the kWh cost price was due to scale effect from MOD.0A to MOD.1 and to technological improvements from MOD.1 to MOD.2.

2. MINIMUM GENERATING COSTS AS A FUNCTION OF THE WIND MACHINE DIAMETER AND OF THE MEAN AND RATED WIND SPEED

Research concerning the variability of the cost price per kWh has been carried out in several countries, especially in the U.S.A., Denmark, the Netherlands and Great Britain.

The latest investigations show that the kWh cost price diminishes when the machine diameter increases.

At equal rated wind speed, the generating cost also decreases when the mean wind speed increases.   This is quite understandable since the same wind plant can furnish a higher amount of electricity at higher wind velocities without supplementary equipment.

At equal mean wind velocities, the calculations show that the kWh cost price reaches its minimum at a certain rated wind speed.   Above and below this wind velocity, the price is higher.

This optimal rated wind speed varies from about :

$$V_R = 2 \, V \quad \text{mean} \qquad \text{for V mean} = \ 5 \ \text{m/s}$$
to
$$V_R = 1.5 \, V \, \text{mean} \qquad \text{for V mean} = 10 \ \text{m/s}.$$

The exact value must be determined from the meteorological data for every important wind plant.

Fig. 186 gives the variation of the kWh cost price as a function of the mean wind speed at 10 m above the ground surface for the most recent wind installations erected in the U.S.A.

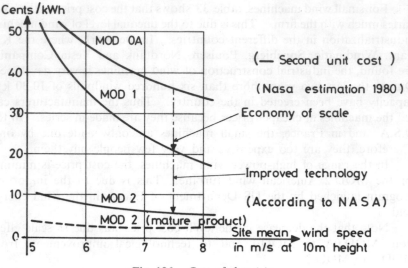

Fig. 186 – *Cost of electricity.*

3. GENERATING COST FOR OTHER ENERGY SYSTEMS

Tables 34 and 35 give the prices of energy produced by petrol and Diesel motor-generators and the prices of electric energy sold by the French national company EDF to its private affiliates. The energy rates are different for domestic, agricultural, professional uses and industry.

Table 36 concerns the cost prices of electricity in France. It shows that the nuclear kWh is the cheapest. Water power and thermal energy (coal and fuel) are more expensive. The kWh obtained in fuel power stations costs about three times the nuclear kWh. The fact that all economical water power sites are already equipped explains the extensive program of construction of nuclear power stations scheduled by the French Government.

Table 37 gives the solar kWh cost price for the French solar power unit of Targasonne (Pyrénées) in which flat reflectors focus solar energy on a boiler placed at the summit of a tower. The solar kWh cost price is very high. It will be hard to reduce it enough to make solar energy competitive for producing mechanical power. Solar energy is more suitable for heating buidings and greenhouses through solar heaters rather than for producing mechanical power through solar thermal power units which are very expensive, except in the case of solar cells. In windy places, large wind rotors connected to electrical grids can supply mechanical power at cheaper prices. The expenditure incurred for the construction of the solar plant of Targasonne is equivalent to the price of five large wind turbines such as the MOD.2 American wind rotor, each of them being able to produce, per year, three times the energy provided by the solar plant !

TABLE 34

Petrol motor-generator (Amortization on one year)

Trade mark	Power in kW	Cost in $	Consumption in l/h	Lifetime in hours	Fuel cost cents/kWh	Capital cost. cents/kWh	Total cents /kWW
Briggs-	0.8	480	0.7	2 000	64	35	99
Stratton	1.3	545	1	2 000	56	25	81
Leroy-	2	780	1.3	3 000	47	15	62
Somer	4	1 200	2.4	3 000	44	12	56

Diesel-generator (Amortization on two years)

Lister ST_1 + Alt. Leroy-Somer	3.5 kW	2 200	1.4	5 000	22	15.6	37.6
Lister ST_2	8	3 600	2.9	8 000	20	7.0	27
Lister ST_3	12.5	7 650	4.5	10 000	20	7.6	27.6
Lister HR_2	16.25	7 650	5.9	10 000	20	5.8	25.8
Lister HR_3	25	8 750	9	12 000	20	2.9	22.9
Lister HR_4	34.4	9 500	12.5	15 000	20	2.3	22.3
Fiat 6 cyl + Alt. Leroy-Somer	50	12 000	18	15 000	20	2	22

TABLE 35 : **Cost of electricity provided to its affiliates by E.D.F.**
A. Low voltage : 220-330 V. (In cents/kWh).
(Rate : August 16, 1980)

1°) *Domestic and agricultural uses*

Single rate	Rate with dead hours		
	full hours		dead hours
6 c	6 c		3.6 c

This price must be increased about 30 % for various taxes + $ 10 to $ 20 per kW installed and per year according to the contract (single rate and rate with dead hours).

2°) *Professional uses*

first part	Supplement	dead hours
10 c	6 c	3.6 c

The first part of the consumption expressed in kWh is equal to 2.5 times
the maximal power installed expressed in kW.

B. Medium voltage : from 380 to 3 000 V. (In cents/kWh).
(General rate)

	Winter		Summer	
peak hours	full hours	dead hours	high hours	dead hours
9.6 c	6 c	3 c	3.5 c	2 c

+ 20 % of taxes
+ 3 % per kW installed and per year

TABLE 36 : **Cost price of electricity in France**

Nuclear kWh	Coal kWh	Fuel kWh	Hydraulic kWh	
3.4 c	6 c	10.5 c	without reservoir	4 c
			with one reservoir	5 to 6 c
			Lake power station	7 to 9 c

TABLE 37 : **Solar energy : French solar power station Thémis
at Targasonne (Pyrénées-orientales) in process of implementation**

Power : 2 300 kW   Cost : $ 25.6 M   Annual production : 3 GWh
Solar kWh cost price with a lifetime estimated at 20 years : $ 1/kWh according to
the French Company E.D.F.

4. ANALYSIS OF THE SITUATION AND CONCLUSIONS

The preceding tables show that windmills can be competitive with
the other means of producing energy in many cases.

In comparaison with previous years, their position in the world energy
market has improved.   This is because petroleum prices have considerably
increased over the last three years, but also because the technology of wind
power systems has greatly improved during this time.

The numerous large wind turbines which have been operating in
Denmark and in the United States for four or five years without serious
problems, as well as the improvements contributed by the Americans with

the construction of MOD. 2 (which has led to an important decrease in the kWh cost price) augur well for the development of wind energy exploitation.

It is beyond doubt that an increase in traditional energy prices would cause further development of the wind machine industry. Construction on a production basis would allow great reductions in the cost of wind machines. To envisage a 50 % reduction in their generating costs is not unrealistic.

However for the time being, in the range of high-power units nuclear energy seems to be the most economic source of energy. But it must be noted that nuclear power stations are only constructed for consumption superior to 100 000 kW. When needs are less, wind energy can be a valuable source of power if the wind is strong enough. This is the case for large islands where wind power stations ranging from 500 kW to 2 000 kW can supply energy competitively. However, as large wind turbines are not sources of regular power, they have to be used in conjunction with a complementary system like Diesel motor-generators or water turbines having a total capacity of at least five or six times the installed wind power. For full utilization of the available wind power, the grid must also be capable of absorbing it at any moment.

In the range of low-power units, the wind generators made by the Danish firms can provide energy at a cheaper price than the petrol or Diesel motor-generators. This is due to the policy of the Danish Government which has favoured the development of wind energy systems. The wind machines made in series can be connected to the grid without any problems. During windy periods, the excess power delivered by the wind generators is given to the grid.

On the other hand, during periods of non-productive winds, the grid supplies energy to the owners of a wind plant. Thus the network acts as a storage reservoir.

The wind plants can also be used autonomously. In places where traditional energy sources are missing or where fuel is not easily supplied (lighthouses for example, small islands and remote countries like the Antarctic continent) but where the wind velocity is high, wind machines even of small or medium sizes constitute an irreplaceable means of producing energy. They are also well suited for providing energy to rural communities in developing countries where the wind velocity is high. In these countries, grids are limited to big towns because of the scattering and low density of the rural population. In this connexion, it must be pointed out that in windy areas, wind motors are much more economical than solar motors. The irregularity of wind power, noted above, is far more acceptable in the rural areas of a developing country, where even interrupted power is a big improvement over no power at all, than in a developed country where people are sensitive to even the shortest power cut.

Therefore, in some parts of the world, the use of wind energy remains a very economic means of supplying power.

Moreover, wind machines must be credited with the following advantages :
— They are silent and non-polluting;
— They are reliable and need only minimal maintenance;
— The energy they use does not pose any supply problem : Wind is free and indefinitely renewable.

APPENDIX

COMPUTER PROGRAMS

The following pages contain six computer programs.

The OPTI programs enable us to calculate the inclination angle and the quantity $C_l bl/r$ relative to the optimal running conditions as a function of the tip-speed ratio. The OPTI 1 program refers to Glauert's theory and the OPTI 2 program to Hutter's theory.

The AERO program allows us to determine the blade chord, the torque and power coefficient of the horizontal-axis wind turbines as a function of the tip-speed ratio.

The EOLE program is planned for calculating the horizontal-axis wind rotor with straight leading and trailing edges. Two or three sections of the blades determined according to the classical methods are chosen, then the leading and trailing edges of the other profiles are aligned with the previous ones. The program allows the torque and power coefficients to be determined as a function of the tip-speed ratio.

The AVEL program enables the torque and power coefficients of an already existing windmill to be calculated from the geometry and the aerodynamic characteristics of the sections of its blades.

The FANN program relates to vertical-axis wind rotors like the Darrieus type with parabolic blades. It may be applied to windmills whose swept area has a cylindrical, conical, conical-truncated, or spherical shape with some adaptations (by introducing the equation giving the shape of the blades in the variables RR and D.).

The ROLL program is especially well adapted to the calculation of slow horizontal axis wind turbines.

The various programs must be used with aerodynamic polars corresponding to wings having an infinite span. However, we have made an exception for the ROLL program for which we have used a polar relative to a wing aspect ratio a = 5. The metallic structure which supports blades of the slow-wind machine actually produces supplementary drags difficult to evaluate. Therefore, the use of a polar corresponding to a wing aspect ratio a = ∞ would not lead to better results.

In the different programs, the geometrical and aerodynamic characteristics of the rotors are designated in FORTRAN by the following expressions :

FORTRAN	Designation
D	D : Diameter
GR	R : Radius
PR(I)	r : Distance to the axis.
RR(I)	r/R
ALFA(I)	α : Setting angle or pitch angle
GI(I)	I : Inclination angle
PI(I)	i : Incidence angle
EL(I)	l : Chord of the section I
EL1(I)	l_1 : Effective chord
X(I) Y(I)	x y $\Big\}$ Coordinates of the leading edge relative to the trailing edge
TGEPS	$\tan \varepsilon = C_d/C_l$
CT	cot I
AK	k
ALO	λ_0 : Nominal tip-speed ratio
ALI	λ : Speed ratio at radius r.
AL(I, J)	λ : Speed ratio at radius r at variable incidences
CM	C_m : Torque coefficient
CP	C_p : Power coefficient
BLA	b : Number of blades
T(I)	tan I
TALFA(I)	tan α
PIP(J)	i : Variable incidence
AM(I, J)	Number proportional to the axial moment due to the element dr for a given incidence
AM1(I, J)	Number proportional to the axial moment due to the different elements dr situated at radius r for an integer value λ_0
C(N)	Number proportional to the torque due to the whole rotor for an integer value λ_0
CL(I) or CL(J)	C_l : Lift coefficient
CD(I) or CD(J)	C_d : Drag coefficient

Particular notations relative to the FANN program

H	Half-height of the rotor
R	Maximal radius
RR	r/R
S	Half of the intercepted area
S	S = 4/3 HR for a parabolic rotor
D	Angle δ
Z	z : Altitude of the element
T	θ : Polar angle
PI	i : Incidence angle in degrees
TGPI	tan i
CT	C_t : Aerodynamic tangential coefficient > 0 towards the leading edge
CN	C_n : Aerodynamic normal coefficient
Wu	W_u
CD0	C_{do} : Drag coefficient in a still atmosphere
X	$\omega R/V$: Speed ratio

Remarks

Each program has been established with a determined aerofoil. If another profile is used, the law of variation of the lift and drag coefficient as a function of the incidence angle corresponding to the new aerofoil must be substituted for the previous one.

Verification of the proposed programs

To prove their validity, we have applied the programs to wind generators already tested in wind tunnels.

The AVEL program and two other programs called AVEL 1 and AVEL 2 derived from this one and from the AERO program, were respectively applied to :

1°) a Soviet four-bladed wind generator having a diameter of 3.60 m designed for $\lambda_0 = 3$ with a Zhukovsky aerofoil. The results of the experiments are seen on pages 151-153 of "Wind powered machines", a book written by Ivan Shefter.

2°) a two-bladed wind rotor of 1.68 m diameter, designed for $\lambda_0 = 7$ with a NACA 4414 aerofoil and described by G. Klein in the report LTR.LA 183 of the National Research Council of Canada.

3°) a model reproducing the Nogent-Le-Roi wind generator on a scale 1/20, having a diameter of 1.555 m, tested by L. Romani at the Eiffel Laboratory in Paris.

The FANN program was applied to the three-bladed Darrieus rotor studied by R. J. Templin and P. South in Toronto and whose test results are included in the conference proceedings of Albuquerque, New Mexico, in May 1976.

Figure 182 shows that the coincidence between the experimental points and the computed points is excellent for the first wind generator having a slow tip-speed ratio ($\lambda_0 = 3$) and for the three-bladed Darrieus rotor.

For the model of the Nogent-Le-Roi wind generator, the coincidence is not so good. At low-speed ratios, the results differ. The actual torque is less than the computed torque. This divergence can be explained by the separation of air flow at blade which is accompanied by the creation of oscillating wakes and therefore by increasing power losses when tip-speed ratio is low.

It must be noted that when applying the AVEL program to rotors running at high tip-speed ratio, negative values of the parameter k may be obtained over the nominal tip-speed ratio λ_0. The blade element concerned works as a fan. When this state extends over a large portion of the blade radius, it will then result in a backflow against the direction of the wind, and thus in the formation of standing annular vortices. As a further consequence of this phenomenon, the power coefficient passes through zero and finally becomes negative.

Although not intended to replace the wind tunnel tests, the appended computer programs make it possible to determine wind rotor performances with reasonable accuracy.

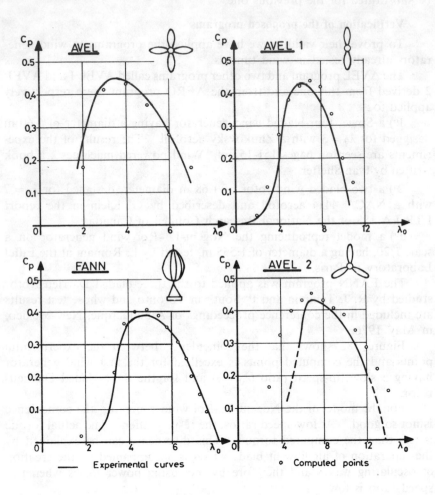

Fig. 187 – *Application of AVEL and FANN program to experimental studies.*

OPTI 1 PROGRAM
(GLAUERT)

```
      DO 1 I=1,100
      AI = I
      AL= AI * 0·1
      PPI = 3·14159
      ALI = ALO * PR (I)
      TETA = PPI / 3·+(ATAN ( ALI )) / 3.
      AK =(COS(TETA)) * SQRT ( 1. + ALI * *2 )
      AH=SQRT (1. + (1·-AK**2)/(AL**2))
      CP =(1 +AK)*(AH-1·)*AL**2
      AL1 = SQRT((1. -AK**2)/(AH**2-1.))
      ALE = AL * (1·+AH) / (1·+AK)
      GIO = 180· *ATAN (1./ALE )/PPI
      CLBR= 8·* PPI *(1·-AK)/((1+AK)*ALE *SQRT(ALE**2+1))
1     WRITE  (6,2) AL,ALE,AK, AH , CP ,CLBR ,GIO
2     FORMAT ( 2X , 13F7.3 )
      STOP
      END
```

OPTI 2 PROGRAM
(HÜTTER)

```
      DIMENSION   AK (100), AH (100),CP (100 )
      CP (1) = 0·
      PPI = 3·14159
      DO 1 L = 1,100
      AL=0·1 * L
      DO 2 I = 2,30
      AI = I
      AK (I) = 0·20 + 0·01 * AI
      AH (I) = SQRT (1.+2.* AK(I)* (1.-AK(I))/(AL**2))
      ALE=AL* (1·+AH(I)) / (1·+AK (I))
      CP (I) =(AL**2 * ((1. + AK(I))**2)*(AH(I) -1.)/(2·*AK(I))
      IF (CP(I) -CP(I -1)) 3,2,2
3     AKO = AK (I -1)
      AHO = AH (I -1)
      CPM = CP (I -1)
      ALE = AL* (1·+AHO) / (1·+AKO )
      CLBR = 8·*PPI * (1. -AKO ) / ( (1. +AKO)* ALE *SQRT (ALE**2+1·))
      GIO = 180· * ATAN (1. /ALE )/ PPI
      WRITE (3,4) AL, ALE ,AKO,AHO, CPM , CLBR , GIO
4     FORMAT (2X , 10 F 7.3 )
      GO TO 1
2     CONTINUE
1     CONTINUE
      STOP
      END
```

AERO PROGRAM

```
C    FAST  WIND  TURBINE (H·A·) NACA  23015  AEROFOIL
     DIMENSION  CL(40), CD(40), PR(10), RR(10), EL(10), T(10),
    1PI(10), GI(40), ALFA(10), PIP(40), C(20), AL(10,40),
    1 AF(10,40), AF1(10,40), AM(10,40), AM1(10,40)
     PPI = 3·141593
     CO = 180·/ PPI
     GR = 5·
     BLA = 3·
     ALO = 7·
     CTE = 8·* PPI / BLA
     DO 2 I = 1, 10
     AI = I
     RR (I)= 0·1 * AI
     PR (I)= RR(I)* GR
     ALI = ALO * RR (I)
     TETA = PPI / 3·+(ATAN (ALI))/ 3·
     AK =(COS(TETA)) * SQRT (1 + ALI * * 2 )
     AH = SQRT (1·+(1·-AK**2)/(ALI * * 2
     ALER = ALI * (1· +AH)/(1·+AK)
     T (I)=1·/ ALER
     GIR  =  ATAN (T (I))
     GI (I) = GIR * CO
     PI (I) = -0·75 * AI +12·75
     ALFA (I) = GI(I) - PI(I)
     CL (I)= 0·1 + 0·11 * PI (I)
     IF (PI(I) -10·) 16 , 16 ,17
17   CL (I)= 1·5 -0·0188 * (PI (I) -14·)**2
16   EL (I)= CTE*PR(I)*(1·-AK )/((1·+AK)* ALER *SQRT(ALER**2+1·)*CL(I))
2    WRITE (3,3) I, ALFA (I), PR (I), RR (I), GI (I), PI (I), CL (I), EL (I)
3    FORMAT (2X , I2 , 10F7·3 )
C    AERODYNAMIC  CHARACTERISTICS
     DO  6  I = 1, 10
     DO  6  J = 1, 40
     L = 40 -J +1
     BL = L
     CT = BL * RR (I)
     GIR = ATAN (1·/ CT )
     GI (J) = GIR * CO
     PIP (J)= GI (J) -ALFA (I)
     IF  (PIP (J) -10·) 4,4,2
4    CL (J)= 0·1 +0·11 * PIP (J)
     CD (J)= 0·007+0·0055 *(CL(J) -0·2)**2
     GO  TO 9
5    IF  (PIP (J) -19· ) 20,20,21
20   CL (J)= 1·5 - 0·0188 * (PIP (J) -14·)**2
     CD (J)= 0·0125 + 0·16 *(CL (J) -1·2)**2
     GO   TO 9
21   CL (J)= 1·
     CD (J)= 0·0125 + 0·16 * (CL(J) -1·2)**2
9    TGEPS  = CD (J)/ CL (J)
     G = CL (J)*BLA* EL(I)*(CT +TGEPS)* SQRT (1+CT**2)/(8·* PPI * PR (I))
     AK = (1·-G)/(1·+G)
     E = CL (J)* BLA * EL(I)*(1·/ CT -TGEPS)* SQRT (1·+CT**2)/(8·*PPI * PR (I))
     AH =(1·+ E)/(1·-E)
     AL (I,J) = (1·+AK)* CT /(RR(I)* (1·+AH))
     AF (I,J) = (1·-AK**2)* RR (I)
     AM (I,J)= RR(I)* RR (I) * ((1·+AK )* * 2)* E * CT
```

AERO CONTINUATION

```
      IF((AL(I,J))*(AL(I,J)-30.))  19,6,6
19    WRITE(3,7)I,J,PIP(J),GI(J),AK,AL(I,J),AM(I,J),G,E,AH,AF(I,J),
     1CZ(J),TGEPS,CT,PR(I),EL(I)
6     CONTINUE
7     FORMAT (2X,I2,2X,I2,2X,8F10.5,/,2X,8F9.4)
      DO 15  I=1,10
      DO 15  N=1,20
      AF1(I,N)=0.
15    AM1(I,N)=0.
      DO 8  I=1,10
      DO 8  J=1,39
      IF((AL(I,J))*(AL(I,J)-21.))  18,8,8
18    N= AL(I,J)
      N1 = AL(I,J+1)
      IF  (N-N1)8,8,10
10    AM1(I,N)=AM(I,J)+(AM(I,J+1)-AM(I,J))*
     1(N-AL(I,J))/(AL(I,J+1)-AL(I,J))
      AF1(I,N)= AF(I,J)+(AF(I,J+1)-AF(I,J))*
     1(N-AL(I,J))/(AL(I,J+1)-AL(I,J))
      WRITE (3,14)I,N,AM1(I,N),AF1(I,N)
14    FORMAT (2I4,2F10.5)
      N=N-1
      IF (N-N1) 8,8,10
8     CONTINUE
      DO 12  N=1,20
      CF=AF1(1,N)*(RR(2)-RR(1)) +AF1(10,N)*(1.-RR(9))
      C(N)=AM1(1,N)*(RR(2)-RR(1)) + AM1(10,N)*(1.-RR(9))
      DO 13  I=2,9
      CF=CF+AF1(I,N)*(RR(I+1)-RR(I-1))
      C(N)= C(N)+AM1(I,N)*(RR(I+1)-RR(I-1))
13    CONTINUE
      ALO=N
      ETAP=(1.-0.93/(BLA*SQRT(ALO**2+0.445)))**2
      CM = C(N)*ETAP
      CP=ALO *CM
12    WRITE (3,11) ALO,CF,CM,CP
11    FORMAT (2X,4F10.5)
      STOP
      END
```

EOLE PROGRAM

```
      (FAST WIND  TURBINE  WITH STRAIGHT LEADING AND  TRAILING  EDGES )
C     GEOMETRICAL DETERMINATION ( N A C A  23015 AEROFOIL )
      DIMENSION  T(10),CL(40),CD(40),PR(10),X(10),Y(10),AF(10,40),
     1RR(10),PI (10),GI(40),EL(10),EL1(10),TALFA (10),AF1(10,40),
     1PIP(40),AM(10,40),AM1(10,40),AL (10,40),ALFA(10),C(20)
      GR=5.
      ALO = 7.
      BLA = 3.
      PPI = 3.141593
      CO =180./PPI
      CTE= 8.*PPI / BLA
      I1= 2
      I2= 5
      I3= 9
      DO 2  I=1,10
      AI = I
      RR(I)=0.1*AI
      PR(I)=0.1*AI*GR
      ALI =ALO*RR(I)
      TETA =PPI/3. +(ATAN(ALI)) /3.
      AK =(COS(TETA)) * SQRT(1.+ALI**2)
      AH=SQRT (1.+(1.-AK**2) /(ALI**2))
      ALER =ALI* (1.+AH)/ (1.+AK)
      T(I)=1. /ALER
      GIR=ATAN(T(I))
      GI (I)=GIR*CO
      PI (I)= -0.75*AI +12.75
      ALFA (I)=GI (I)-PI (I)
      ALFAR =ALFA (I)/CO
      IF (PI (I)-10.) 6,6,7
  6   CL (I)=0.1+0.11*PI (I)
      GO TO 8
  7   CL (I)=1.5- 0.0188*(PI (I)-14.)**2
  8   EL (I)= CTE*PR(I)* (1.-AK)/ ((1.+AK)*ALER *SQRT (ALER**2+1.)*CL (I))
      X (I)=EL (I)*COS(ALFAR)
      Y (I)=EL (I)*SIN (ALFAR)
  2   WRITE (3,5)I,GI (I),ALFA (I),PI (I),EL (I),PR (I),X (I),Y (I)
  5   FORMAT (2X,I2,2X,9F 7.3)
      DO 9  I=1,10
      IF  (I-I2)10,10,11
 10   X (I)= X (I1)+(X (I2)-X (I1))*(PR(I)-PR(I1))/(PR(I2)-PR(I1))
      Y (I)= Y (I1)+(Y (I2)-Y (I1))*(PR(I)-PR(I1))/(PR(I2)-PR(I1))
      GO TO 12
 11   X (I)=X (I2)+(X (I3)-X (I2))*(PR(I)-PR (I2))/(PR(I3)-PR(I2))
      Y (I)=Y (I2)+(Y (I3)-Y (I2))*(PR(I)-PR(I2))/(PR(I3)-PR(I2))
      TALFA (I)=Y (I)/X (I)
      ALFAR =ATAN (TALFA (I))
      ALFA (I)=ALFAR*CO
      PI (I)=GI (I)-ALFA (I)
 12   EL1 (I)= SQRT (X (I)**2+Y (I)**2)
  9   WRITE (3,5)I,GI (I),ALFA (I),PI (I),CL (I),PR(I),EL (I),X (I),Y (I),EL1 (I)
C     AERODYNAMIC  CHARACTERISTICS
      DO 13  I = 1,10
      DO 13  J = 1,40
      L = 40 -J +1
```

EOLE CONTINUATION

```
       BL = L
       CT= BL * RR( I )
       GIR=ATAN (1. / CT )
       GI ( J )= GIR * CO
       PIP ( J )= GI ( J ) -ALFA ( I )
       IF  (PIP ( J )-10. )14,14,15
14     CL(J)= 0.1+0.11 * PIP ( J )
       CD( J )=0.007 +0.0055 * (CL( J ) -0.2)**2
       GO TO 16
15     IF (PIP (J )-19. ) 26,27,27
26     CL( J )=1.5-0.0188 * (PIP (J ) -14. )**2
       CD(J)=0.0125+0.16 * (CL (J ) -1.2)**2
       GO TO 16
27     CL( J ) = 1.
       CD(J) = 0.0125+0.16 * (CL (J )- 1.2)**2
16     TGEPS= CD(J)/CL(J)
       G=CL(J)*BLA* EL1(I) * (CT+TGEPS)* SQRT(1+CT**2) / ( 8.* PPI*PR(I))
       AK=(1-G)/(1+G)
       E=CL(J)*BLA* EL1( I)* ( 1. /CT-TGEPS)*SQRT(1. +CT**2)/( 8.*PPI*PR( I))
       AH=(1.+E)/(1.-E )
       AL( I,J )= (1.+AK)* CT/( RR(I) * ( 1.+AH))
       AF( I,J )= (1.-AK**2)* RR(I)
       AM( I,J )= RR(I)* RR(I) * ( ( 1 +AK)**2 )* E * CT
       IF ( (AL ( I,J ) )* (AL ( I,J )-30) ) 29,13,13
29     WRITE (3,24) I,J,PIP ( J ),GI ( J ),AK,AL(I,J) ,AM(I,J) ,G,E,AH,AF( I, J),
      1 CL(J),CD(J),TGEPS,CT, PR( I ),EL1( I)
13     CONTINUE
24     FORMAT (2X,I2,2X,I2,2X,9F10.5,/,2X,6F10.5 )
       DO 28 I=1,10
       DO 28 N= 1,20
       AF1 (I,N)= 0.
28     AM1 (I,N)= 0.
       DO 17 I = 1,10
       DO 18 J = 1,39
       IF ((AL(I,J))* (AL(I,J) -21. )) 30,18,18
30     N=AL (I,J)
       N1= AL( I,J +1)
       IF  (N-N1) 18,18,19
19     AF1 (I,N)=AF(I,J)+(AF(I,J+1) -AF(I,J)) *
      1 (N-AL(I,J))/(AL(I,J+1) -AL(I,J))
       AM1 (I,N)= AM(I,J)+(AM (I,J+1)-AM(I,J))*
      1 (N-AL (I,J))/(AL (I,J+1)-AL (I,J))
       WRITE (3,20) I,N,AM1(I,N),AF1 (I,N)
20     FORMAT (2I4, 2F12.6 )
       N=N-1
       IF (N) 17,17,25
25     IF(N-N1) 18,18,19
18     CONTINUE
17     CONTINUE
       DO 21 N = 1,20
       CF=AF1(1 N)* (RR(2)- RR(1))+AF1 (10 N)* (1.- RR(9))
       C(N)=AM1 (1,N) * (RR(2)- RR(1)) +AM1(10,N)* (1.- RR (9))
       DO 22 I = 2,9
       CF = CF +AF1(I N)* (RR(I +1)- RR(I -1))
       C(N)= C(N)+AM1(I,N)* (RR(I +1)- RR(I-1))
22     CONTINUE
       ALO = N
       ETAP =(1.- 0.93/(BLA * SQRT ( ALO**2 +0.445)))**2
       CM= C(N) * ETAP
       CP = ALO * CM
21     WRITE (3,23) ALO , CF, CM , CP
23     FORMAT ( 4F10.4 )
       STOP
       END
```

AVEL PROGRAM

```
C     GEOMTRICAL DESCRIPTION   JOUKOVSKY AEROFOIL
      DIMENSION CL(40),PR(10),RR(10),GI(40),EL1 (10),
     1 PIP (20 ),AM(10,20),AM1 (10,40),AL(10,20),ALFA(10),C(20)
      READ(2,1)(EL1(I),I=2,10,2),(ALFA(I),I=2,10,2)
   1  FORMAT(2X,10F5.2)
      GR=1.8
      PPI=3.141593
      CO=180./PPI
      BLA=4.
      DO 2  I=2,10,2
      AI=I
      RR(I)=0.1*AI
      PR(I)=0.1*AI*GR
   2  WRITE(3,3)I,RR(I),PR(I),EL1(I),ALFA(I),GR
   3  FORMAT(2X,I2,5F5.2)
C     AERODYNAMIC CHARACTERISTICS
      DO 6 I=2,10,2
      DO 6 J=1,20
      L=20-J+1
      BL=L
      CT=BL*RR(I)
      GIR=ATAN(1./CT)
      GI(J)=GIR*CO
      PIP(J)=GI(J)-ALFA(I)
      IF(PIP(J)-14.) 18,19,19
  18  CL(J)=0.61+0.09*PIP(J)
      IF(CL(J)-1.)5,5,4
   4  CL(J)=1.13-0.006*(PIP(J)-9.)**2
      GO TO 5
  19  CL(J)=1.
   5  TGEPS=0.016+0.00113*(PIP(J)-1.8)**2
      G=CL(J)*BLA*EL1(I)*CT+TGEPS)*SQRT(1+CT**2)/(8.*PPI*PR(I))
      IF(G+1.)6,6,16
  16  AK=(1-G)/(1+G)
      E=CL(J)*BLA*EL1(I)*(1./CT-TGEPS)*SQRT(1.+CT**2)/(8.*PPI*PR(I))
      AH=(1.+E)/(1.-E)
      AL(I,J)=(1.+AK)*CT/(RR(I)*(1.+AH))
      AM(I,J)=RR(I)*RR(I)*((1.+AK)**2)*E*CT
      IF(AL(I,J)-21.) 17,6,6
  17  WRITE(3,7)I,J,PIP(J),GI(J),AK,AL(I,J),AM(I,J),G,AH,
     1 CL(J),TGEPS,CT,PR(I),EL1(I)
   6  CONTINUE
   7  FORMAT(2X,I2,2X,I2,2X,7F10.5,/,2X,6F10.5)
      DO 15 I=1,10
      DO 15 N=1,20
  15  AM1(I,N)=0.
      DO 8 I=2,10,2
      DO 8 J=1,19
      IF(AL(I,J)*(AL(I,J)-21.)) 9,8,8
   9  N=AL(I,J)
      N1=AL(I,J+1)
      IF(N-N1) 8,8,10
  10  AM1(I,N)=AM(I,J)+(AM(I,J+1)-AM(I,J))*
     1 (N-AL(I,J))/(AL(I,J+1)-AL(I,J))
      WRITE(3,14)I,N,AM1(I,N)
  14  FORMAT(2X,I2,2X,I2,2X,F10.4)
```

AVEL CONTINUATION

```
      N=N-1
      IF (N-N1)8,8,10
   8  CONTINUE
      DO  12  N=1,10
      C(N)= 0.5*(AM1(2,N)+AM1(10,N))*0.4
      DO  13  I=4,8,2
      C(N)=C(N)+AM1(I,N)*0.4
  13  CONTINUE
      ALO = N
      ETAP=(1.-0.93/(     *SQRT(ALO**2+0.445)))**2
      CM=C(N)*ETAP
      CP=CM *ALO
  12  WRITE (3,11) ALO,CM,CP
  11  FORMAT(2X,3F10.5)
      STOP
      END
```

LOAD

0.40 0.40 0.37 0.33 0.3 33.9 23.9 18.3 14.1 11.2

ROLL PROGRAM

```
C       SLOW WIND TURBINE (CURVED THIN PLATE f/l = 0.1)
        DIMENSION T(10),CL(20),CD(20),PR(10),X(10),Y(10),
       1 RR(10),PI(10),GI(20),EL(10),EL1(10),TALFA(10),
       1 PIP(20),AM(10,20),AM1(10,40),AL(10,20),ALFA(10),C(20)
        GR = 2.6
        ALO = 1.5
        BLA = 12.
        PPI = 3.141593
        CO = 180./PPI
        CTE = 8.*PPI / BLA
        I1 = 2
        I2 = 4
        I3 = 8
        DO 2 I = 1,10
        AI = I
        RR(I) = 0.1*AI
        PR(I) = 0.1*AI*GR
        ALI = ALO*RR(I)
        TETA = PPI/3.+(ATAN(ALI))/3.
        AK = (COS(TETA))*SQRT(1.+ALI**2)
        AK = SQRT(1.+(1.-AK**2)/(ALI**2))
        ALER = ALI*(1.+AH)/(1.+AK)
        T(I) = 1./ALER
        GIR = ATAN(T(I))
        GI(I) = GIR*CO
        PI(I) = -0.5*AI + 8.
        ALFA(I) = GI(I)- PI(I)
        ALFAR = ALFA(I)/CO
        IF (PI(I) - 10.) 6,6,7
6       CL(I) = 0.074* PI(I)+0.62
        GO TO 8
7       CL(I) = 1.36
8       EL(I) = CTE*PR(I)*(1.-AK)/((1.+AK)*ALER*SQRT(ALER**2+1.)*CL(I))
        X(I) = EL(I)*COS(ALFAR)
        Y(I) = EL(I)*SIN(ALFAR)
2       WRITE (6,5) I,GI(I),ALFA(I),PI(I),EL(I),PR(I),X(I),Y(I)
5       FORMAT (2X,I2,2X,9F7.3)
        DO 9 I = 1,10
        IF (I-I2) 10,10,11
10      X(I) = X(I1)+(X(I2)-X(I1))*(PR(I)-PR(I1))/(PR(I2)-PR(I1))
        Y(I) = Y(I1)+(Y(I2)-Y(I1))*(PR(I)-PR(I1))/(PR(I2)-PR(I1))
        GO TO 12
11      X(I) = X(I2)+(X(I3)-X(I2))*(PR(I)-PR(I2))/(PR(I3)-PR(I2))
        Y(I) = Y(I2)+(Y(I3)-Y(I2))*(PR(I)-PR(I2))/(PR(I3)-PR(I2))
        TALFA (I) = Y(I)/X(I)
        ALFAR = ATAN(TALFA(I))
        ALFA(I) = ALFAR*CO
        PI(I) = GI(I)- ALFA(I)
12      EL1(I) = SQRT(X(I)**2+Y(I)**2)
        EL1(1) = 0.
        EL1(2) = 0.
        EL1(3) = 0.
9       WRITE (6,5)I,GI(I),ALFA(I),PI(I),CL(I),PR(I),EL(I),X(I),Y(I),EL1(I)
```

ROLL CONTINUATION

```
       C        AERODYNAMIC CHARACTERISTICS
0051            DO 13  I=1,10
0052            DO 13  J=1,20
0053            AJ=J
0054            PIP(J)=3.*AJ-10.
0055            IF (PIP(J)-10.)14,14,15
0056       14   CL(J)=0.074*PIP(J)+0.62
0057            CD(J)=0.0006*(PIP(J)+2.5)**2+0.062
0058            GO TO 16
0059       15   CL(J)=1.36
0060            CD(J)=0.0006*(PIP(J)+2.5)**2+0.062
0061       16   TGEPS=CD(J)/CL(J)
0062            GI(J)=ALFA(I)+PIP(J)
0063            GIR=GI(J)/CO
0064            CT=COS(GIR)/SIN(GIR)
0065            G=CL(J)*BLA*EL1(I)*(CT+TGEPS)*SQRT(1+CT**2)/(8.*PPI*PR(I))
0066            AK=(1-G)/(1+G)
0067            F=CL(J)*BLA*EL1(I)*(1./CT-TGEPS)*SQRT(1.+CT**2)/(8.*PPI*PR(I))
0068            AH=(1.+F)/(1.-F)
0069            AL(I,J)=(1.*AK)*CT/(RR(1)*(1.+AH))
0070            AM(I,J)=RR(1)*RR(1)*((1.+AK)**2)*F*CT
0071            WRITE (6,24)  I,J,PIP(J),GI(J),AK,AL(I,J),AM(I,J),G,F,AH,
               1CL(J),CD(J),TGEPS,CT,PR(1),EL1(I)
0072       13   CONTINUE
0073       24   FORMAT(2X,I2,2X,I2,2X,8F10.5,/,2X,6F10.5)
0074            DO 17  I=1,10
0075            DO 17  N=1,20
0076       17   AM1(I,N)=0.
0077            DO 26  I=1,10
0078            DO 26  J=1,20
0079       26   AL(I,J)=5.*AL(I,J)
0080            DO 18  I=4,10
0081            DO 18  J=1,19
0082            N=INT(AL(I,J))
0083            NI=INT(AL(I,J+1))
0084            IF (N-N1)18,18,19
0085       19   AM1(I,N)=AM(I,J)+(AM(I,J+1)-AM(I,J))*
               1(N-AL(I,J))/(AL(I,J+1)-AL(I,J))
0086            WRITE (6,20)I,N,AM1(I,N)
0087       20   FORMAT (2X,I2,2X,I2,2X,F10.4)
0088            N=N-1
0089            IF (N-N1)18,18,19
0090       18   CONTINUE
0091            DO 21  N=1,20
0092            C(N)=AM1(4,N)*(RR(5)-RR(4))+AM1(10,N)*(1.-RR(9))
0093            DO 22  I=5,9
0094            C(N)=C(N)+AM1(I,N)*(RR(I+1)-RR(I-1))
0095       22   CONTINUE
0096            ALO = 0.2*N
0097            CP = ALO * C(N)
0098       21   WRITE (6,23)ALO,C(N),CP
0099       23   FORMAT (2X,F4.2,2X,F10.4,2X,F10.4)
0100            STOP
0101            END
```

FANN   PROGRAM

```
C     VERTICAL  AXIS  WIND  TURBINE   N A C A  0012  AEROFOIL
      READ(2,1)  H,R,BLA,S,EL
      WRITE(3,1)  H,R,BLA,S,EL
    1 FORMAT (2X,5F6.3)
      CDO = 0.012
      DZ = H / 5.
      DT= 0.1*3.14159
      DTZ= DT *DZ
      PLS = BLA* EL /( 8.*3.14159 * S)
      DO 2 L = 1,14

      F1= 0.
      F3= 0.
      X= L
      DO 3  I = 1,5
      AI = I
      Z =(AI - 0.5)*DZ
      RR =1. -(Z/H)**2
      D =ATAN (2*Z *R / H**2)
      DO 4   J =1,10
      AJ =J

      T =(AJ-0.5) * DT
      TGPI =(SIN (T)*COS (D))/(X * RR + COS (T))
      PIR = ATAN (TGPI )
      PI = PIR *180. /3.14159
      PI = ABS (PI)
      IF (PI-9.) 9, 9, 10
    9 CN = 0.1 * PI
      CT= 0.00164 * PI **2 -CDO
      GO TO 8
   10 IF ( PI -12.5 ) 11, 11, 12

   11 CN = 1.02
      CT= 0.0145 * PI - CDO
      GO TO 8
   12 IF ( PI - 20.) 13, 13, 14
   13 CN = 0.8
      CT= -0.02 * PI + 0.35 - CDO
      GO TO 8
   14 IF ( PI - 80.) 16, 16, 15
   16 CN = 2.1 - 0.000364 * (80. - PI)**2
      CT= 0.05 + 0.1* SIN((PI -85.)*0.035) - CDO

      GO TO 8
   15 CN = 2.1 - 0.000167* (PI-80.)**2
      CT= 0.05 + 0.1* SIN ((PI -85.)* 0.039) - CDO
    8 WU=(X * RR + COS (T))**2 + (SIN(T) * COS (D))**2
      FK = WU *(CN * SIN (T) - CT *COS (T) / COS (D))
      F1 = F1 + FK
      FP = CT * WU * RR / COS (D)
      F3 = F3 + FP
      DD = COS (D)
      TT = COS (T)
      WRITE (3 , 6) I , J , T , TT , D , DD , CN , CT , WU , FK , FP , F1 , F3 , PI
    6 FORMAT ( 2X , 2I2 , 8F7.3 , / , 5F10.2 )
    4 CONTINUE
    3 CONTINUE
```

FANN CONTINUATION

```
        F2 = F1 * DTZ * PLS
        F4 = F3 * DTZ * X
        WRITE ( 3 , 5 ) L , I , Z , D , DD , F2 , F4
    5   FORMAT ( 2X , 2I2 , 5F7. 3 )
        G = 2. * F2
        AK = (1. - G) / ( 1. + G )
        ALO = X * ( 1. + AK ) / 2
        CP. = PLS * F4 *( 1. + AK) * *3
        CM = CP / ALO
        WRITE ( 3 , 7 ) X , ALO , AK , CM , CP
    7   FORMAT ( 2X , 10F 7. 3 )
    2   CONTINUE
        STOP
        END
        LOAD
        2.750    2.300    3.000    8.430    0.152
```

Eiffel polar of flat and curved aerofoils

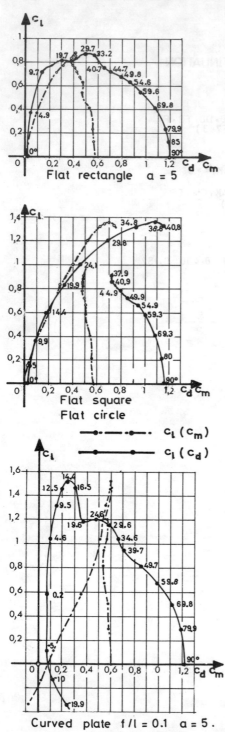

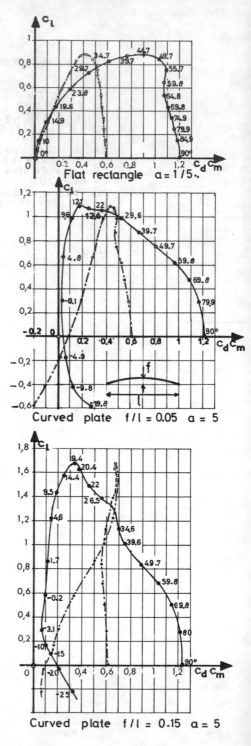

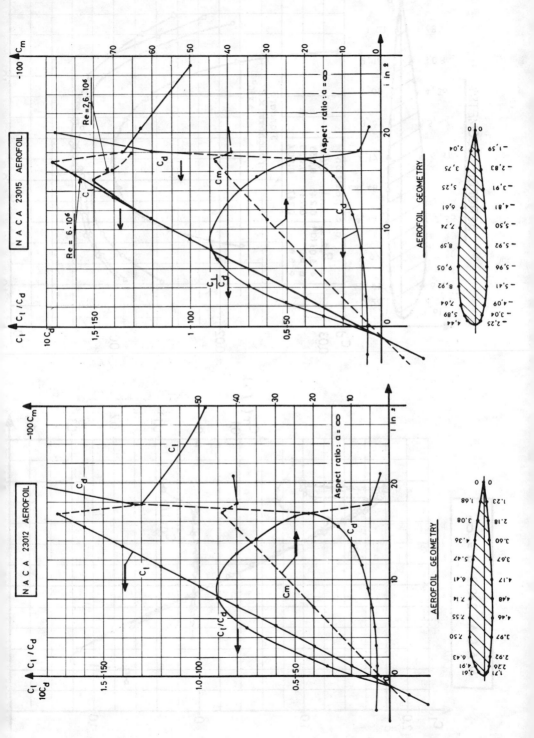

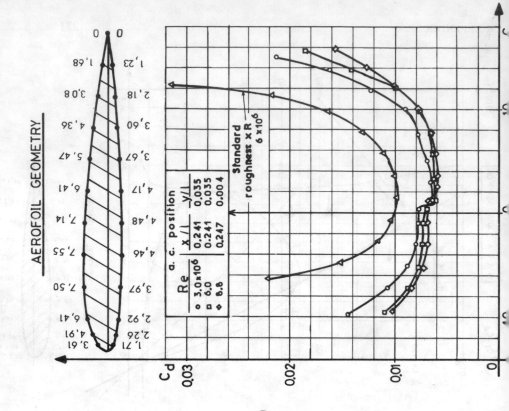

AEROFOIL GEOMETRY

Re	a. c. position	
	x/l	y/l
o 3.0×10⁶	0.241	0.035
□ 6.0	0.241	0.035
◇ 8.8	0.247	0.004

Standard
roughness ×R
6 ×10⁶

C_d

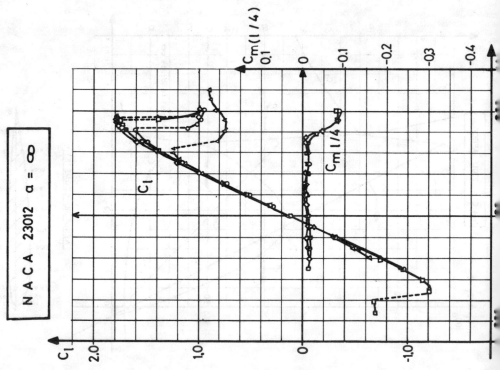

NACA 23012   a = ∞

C_l

$C_m(1/4)$

$C_m \, 1/4$

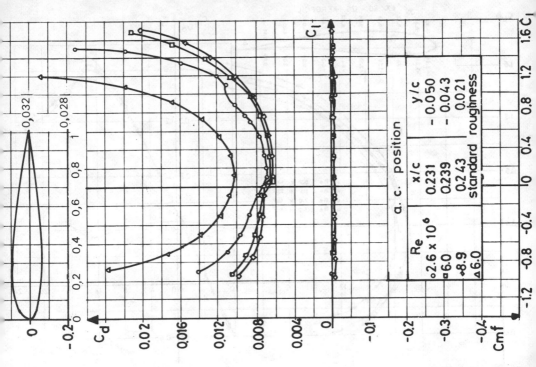

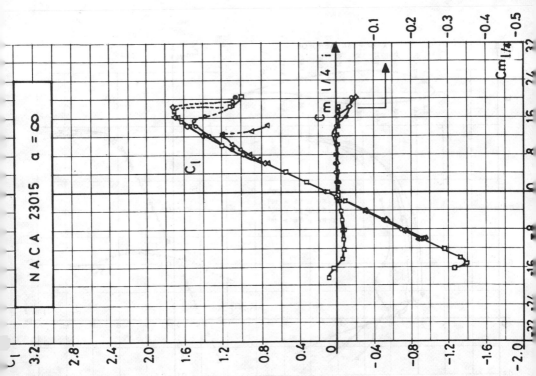

NACA 23015 $a = \infty$

a. c. position

Re	x/c	y/c
o 2.6 × 10⁶	0.231	− 0.050
□ 6.0	0.239	− 0.043
♦ 8.9	0.243	0.021
△ 6.0	standard roughness	

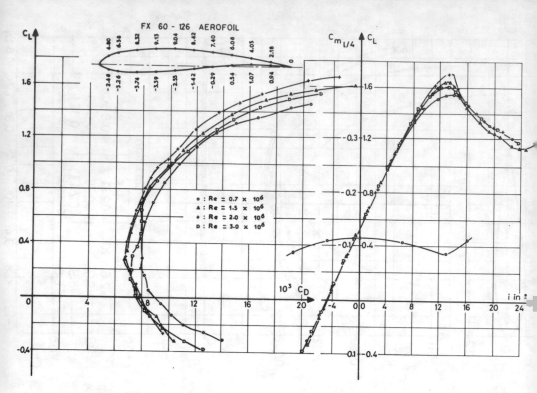

FX 60 - 126 AEROFOIL

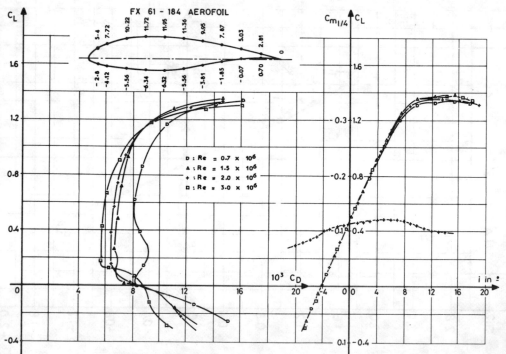

FX 61 - 184 AEROFOIL

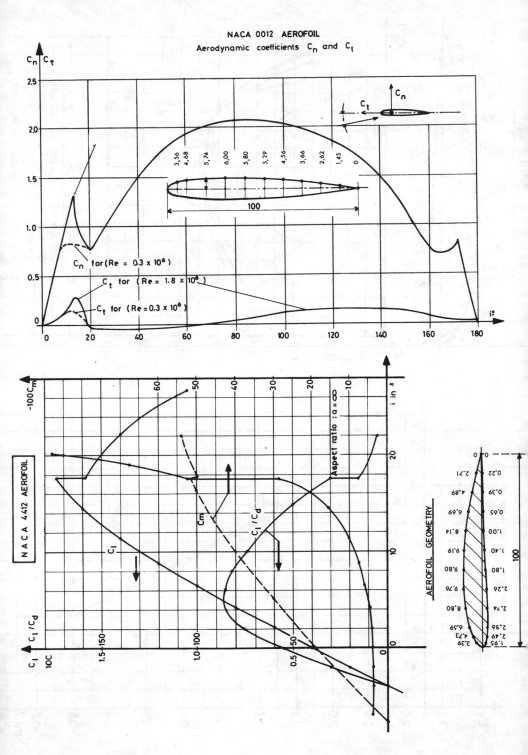

NACA 0012 AEROFOIL
Aerodynamic coefficients C_n and C_t

C_n for (Re = 0.3 x 10⁶)

C_t for (Re = 1.8 x 10⁶)

C_t for (Re = 0.3 x 10⁶)

NACA 4412 AEROFOIL

AEROFOIL GEOMETRY

NACA 4415 a = ∞

AEROFOIL GEOMETRY

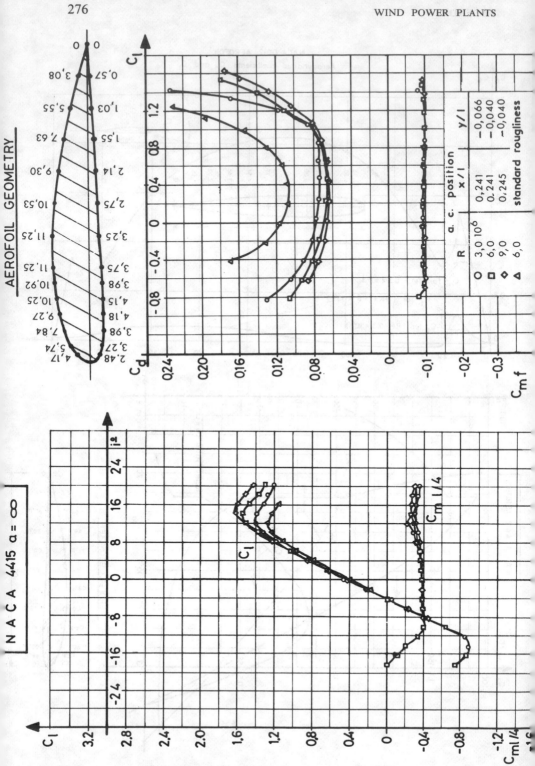

R	a. c. position x/l	y/l
○ 3,0·10⁶	0,241	−0,066
□ 6,0	0,241	−0,040
◇ 9,0	0,245	−0,040
◁ 6,0	standard rougliness	

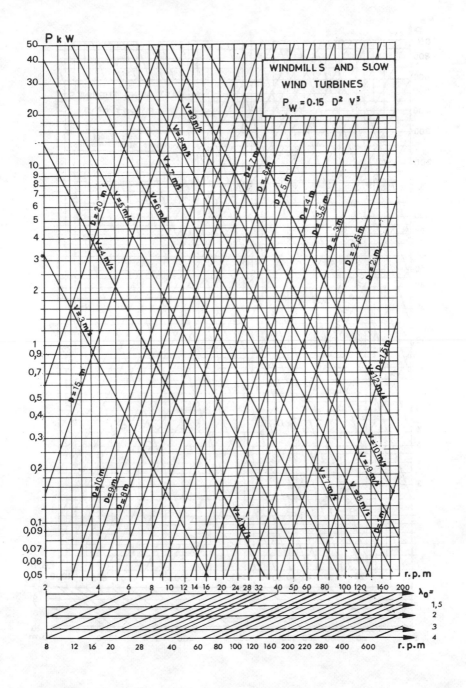

WINDMILLS AND SLOW
WIND TURBINES
$P_W = 0.15 \ D^2 \ V^3$

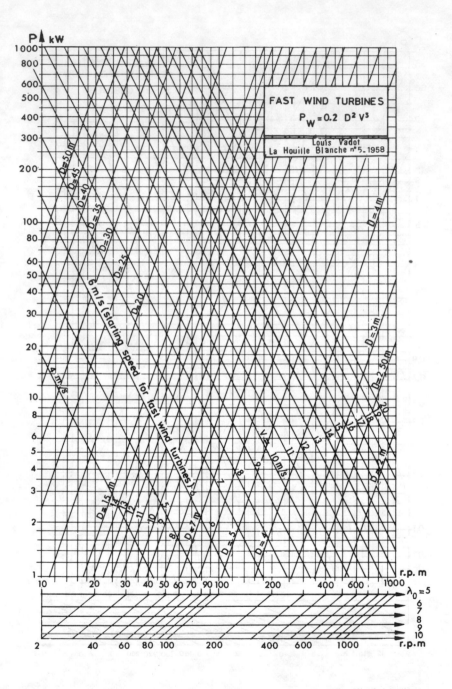

SELECTED BIBLIOGRAPHY

P. AILLERET : *L'énergie éolienne : sa valeur et la prospection des sites.* Revue générale de l'électricité, 1946.

A. BETZ : *Die Windmuhlen im Lichte neuerer Forschung. Die Naturwissenschaft,* Berlin, 1927, Heft 46.

A. BOISSON : *Aérodynamique du vol de l'avion.* Dunod, 1969.

R. BONNEFILLE : *Les réalisations d'électricité de France concernant l'énergie éolienne.* Journal « La Houille Blanche », n⁰ 1, 1975.

BRACE INSTITUTE : *How to construct a cheap wind machine for pumping water* (Savonius rotor).

E. C. CLAVER : *Static and dynamic loadings on the tower of a windmill* (Steering committee for wind energy in developing countries, SWD, Amersfoort, Netherlands).

CHAMPLY : *Moteurs à vent.* Dunod, 1933.

I. CHERET and R. BREMOND : *Étude du régime des Vents en Afrique occidentale — Possibilités d'utilisations des éoliennes pour l'exhaure de l'eau* — Service de l'Hydraulique de l'A.O.F., 1962.

R. CHILCOTT : *Notes on the development of the Brace Airscrew Windmill as a prime mover.* Brace Institute, Mc Gill University, Montréal, 1969.

R. COMOLET : *Mécanique des fluides expérimentale.* Masson, 1969.

F. da MATHA SANT'ANNA : *Les Moulins à vent et l'énergie de demain.* École Polytechnique de Montréal, Canada, 1975.

Th. A. H. DEKKER : *Performance characteristics of some sail and steel-bladed wind rotors.* Amersfoort, Netherlands, 1977.

Ph. DUCHENE and MARRULLAZ : *Distributions statistiques et cartographie des vitesses moyennes de vent en France* (CSTB, Nantes, France, 1977).

DUREMBER, FAVRIS, LAGRANDE, SARRE and F. VAIREL : *Énergie éolienne.* Journal « Écologie », Montargis, France, 1976.

L. ESCANDE : *Cours de Mécanique des fluides.* 1949.

Frank B. ELDRIDGE : *Wind machines.* National Science Foundation Research. Washington, 1975.

H. GLAUERT : " *The elements of airfoil and airscrew theory*". Cambridge University Press, 1959.

E. W. GOLDING : *The Generation of Electricity by Wind Power*. E. and F. Spon Ltd, London, 1976.

HOUGTON and BOCK : *Aerodynamics for Engineering Students*. Butter and Tanner, London MG0.

U. HUTTER : *Planning and balancing of energy of small output wind power plant* (Symposium of New Delhi, 1954).

W. A. M. JANSENS : *Horizontal axis fast running wind turbines for developing countries*. SWD. PO. BOX 85, Amersfoort, Netherlands, 1976.

W. A. M. JANSENS and P. T. SMULDERS : *Rotor design for horizontal axis windmills*. SWD, Amersfoort, 1977.

J. JUUL : *Wind Machines*. Symposium of New Delhi, 1954.

G. J. KLEIN : The design of high speed windmills suitable for driving electric generators. Report LTR – LA 183, National Research Council Canada NRC, 1975.

U. KRABBE and O. H. GLESEN : *Electronic control of voltage and frequency of windmills not connected to the grid* (University of Copenhagen).

G. LACROIX : *Les éoliennes électriques Darrieus*. Journal « La Nature », 1929.

G. LACROIX : *L'énergie du vent*. Journal « la Technique Moderne », 1949.

H. LANOIX : *Les aéromoteurs modernes*. Librairie Girardot, 1947.

Paul L. LEFEBVRE and Duane E. CROMACK : *A comparative study of optimized blade configurations for high speed wind turbines*. Technical Report prepared for the U.S. Energy Research and Development Administration by the Mechanical Engineering Department of Amherst, Massachussetts University, 1977.

P. LUNDSAGER and GUNNESKOW : *Static deflection and eigenfrequency. Analysis of the Nibe wind turbine rotors*. Theoretical background. (Riso, 1980).

P. LUNDSAGER, FRANDSEN, CHRISTENSEN : *Analysis of data from the Gedser wind turbine 1977-1979* (Riso, 1980).

LYSEN, BOS and CORDES : *Savonius rotors for waterpumping*. SWD, Amersfoort, Netherlands).

J. MARTIN : *L'énergie éolienne demain*. Journal Arts et Métiers, 1975.

J. NOUGARO : *Cours d'aérodynamique*. ENSEIHT (Toulouse).

OUZIAUX and PERRIER : *Fluides compressibles Aérodynamique*. Dunod, 1967.

P. C. PUTNAM : *Power from the Wind*. Van Nostrand Reinhold Company, New York, 1947.

Jack PARK : *Simplified systems for experimenters*. Helion, 1976.

Helge PETERSEN : *The test plant for and a survey of small Danish windmills* (Riso laboratory, DK 4000, Roskilde, Denmark, 1980).

Helge PETERSEN : *The 12 m long wind turbine blade manufactured by Volund and Boats* (Riso, oct. 1979).

Helge PETERSEN : *Rotorkonstruction for de to Nibe Windmoller opfort of Elvaerkerne* (Riso, 1979).

G. Kh. SABININ : *The ideal windmill* (in Russian), Moscow, 1927.

SERRANE : *Cours d'aérodynamique*. Dunod.

Y. I. SHEFTER : *Wind Powered Machines*. Leo Kanner Associates, Redwood City California, 1974.

H. STEFANIAK : *Windrad grosster Leistungabgabe*. Forschung 19, Bd Heft 1.

P. SOUTH and R. S. RANGI : *An experimental investigation of a 12 ft diameter high-speed vertical-axis wind turbine*. Report LRT – LA 166 NRC, Canada.

R. J. TEMPLIN : *Aerodynamic performance theory for the NRC vertical axis wind turbine*. Report LTR LA 160, NRC, Canada, 1974.

R. J. TEMPLIN : *An estimate of the interaction of windmills in widespread arrays*. Report LTR – LA 171, NRC, 1974.

L. VADOT : *Le pompage de l'eau par éolienne*. Journal « La Houille Blanche », n° 4, 1957.

L. VADOT : *La production d'énergie électrique par éolienne*. « La Houille Blanche », n° 5.

L. VADOT : *La production d'énergie par éoliennes*. Problèmes économiques. « La Houille Blanche », n° 1, 1959.

Von O. FLACHSBART : *Messungen auf ebenen und gewolbten Platten*. Göttingen, 1931.

D. F. WARNE and P. C. CALNAN : *Generation of electricity from the wind*. Proceedings IEE, vol. 124, nov. 1977.

GKSS (Forschungszentrun Geestacht GMBH): *Versuchsfeld Pellworm fur Windkraftanlagen*, 1980.

WEGLEY and Al. : *A Siting Handbook for small energy conversion systems*. Battelle institute, Richland, 1978.

H. MITSCHEL : *Die Grosswindanlage GROWIAN eine moderne Windenergieanlage in Nord Deutschland*. Forschung in der Kraftwerkstechnik, 1980.

FRIES, BORCHERS and PETERSEN : *Comparative investigations on the performance of small wind energy conversion systems*. GKSS Geestacht GMBH, 1980.

Luise JUNGE : *Windkraft Journal*. 2330 Eckernforde, Germany, 1981.

Jos Van BRIEL : *New aerodynamic theory of windmills*. Peer, Belgium, 1981 (in Press).

PROCEEDINGS OF THE CONFERENCE ON NEW SOURCES OF ENERGY, ROME, AUGUST 1961.

ARGAND : *Mesure des paramètres caractéristiques de l'énergie éolienne en vue du choix des sites favorables à l'installation d'aéromoteurs*.

CAMBILARGIU : *Génératrices à aéromoteurs*.

F. DELAFOND : *Essais de l'aérogénérateur. Andreau — Enfield de Grand Vent*.

FRENKIEL : *Wind Flow over hill in relation to wind power utilization*.

A. HAVINGA : *Windmills in Holland*.

V. HUTTER : *The aerodynamic layout of wing blades of wind turbines with igh tip-speed ratio*.

J. JUUL : *Design of wind Power Plant in Denmark*.

J. JUUL : *Recent developments and potentiel improvements in Wind Power utilization for use connection with electrical networks in Denmark. Economy and operation* of wind *power plants.*

A. LEDACS KISS : *Wind Power Plants suitable for use in the National Power supply network in Hungary.*

MORIYA and TOMASAWA : *Wind turbines in Japan.*

MORRISON : *The testing of a wind Power Plant.*

STERNE and FRAGOLANTS : *A wind driven electrical generator coupled into AC network.* The matching problem.

L. VADOT : *Plans et essais d'installations éoliennes.*

VENKITESH-WARAN : *Operation of Allgaier Type (6-8 kW) wind electric generator at Porbandar,* India.

F. VILLINGER : *Small wind electric Plant with permanent magnetic generator.*

J. WALKER : *Utilization of random power with particular reference to small size wind power plants.*

PROCEEDINGS OF THE FIRST INTERNATIONAL SYMPOSIUM ON WIND ENERGY SYSTEMS, CAMBRIDGE, UK, SEPTEMBER, 1976.

A. C. BAXTER : *A low cost windmill rotor.*

B. F. BLACKWELL : *Status of the ERDA/Sandia 27 m Darrieus turbine design.*

T. E. BASE and L. J. RUSSEL : *Computer aided aerogenerator analysis and performance.*

G. I. FEKETE : *A self-contained 5,000 kW capacity wind energy conversion system with storage.*

P. HIRST and D. H. REES : *The regulation, storage and conversion of wind produced electrical energy at the level of a few hundred watts.*

M. J. HOLGATE : *A crossflow wind turbine.*

O. HOLME : *A contribution to the aerodynamic theory of the vertical-axis wind turbine.*

P. B. S. LISSAMAN : *General performance theory for crosswind axis turbines.*

J. P. MOLLY : *Balancing power supply from wind energy converting systems.*

P. J. MUSGROVE : *The variable geometry vertical axis windmill.*

M. E. PARKES and F. J. M. van de LAAK : *Windpower installations for water pumping in developing countries.*

P. M. STORZA : *Vortex augmentors for wind energy conversion.*

J. M. TRICKLAND : *A performance prediction model for the Darrieus turbine.*

N. V. C. SWAMY and A. A. FRITZSCHE : *Aerodynamic studies on vertical-axis wind turbines.*

R. J. TEMPLIN and P. SOUTH : *Some design aspects of high-speed vertical-axis wind turbines.*

Th. van HOLTEN : *Windmills with diffuser effect induced by small tipwanes.*

J. T. YEN : *Tornado-type wind energy system : Basic consideration.*

SECOND INTERNATIONAL SYMPOSIUM ON WIND ENERGY SYSTEMS, AMSTERDAM, OCTOBER 1978.

R. BRAASCH : *On the vertical axis wind turbine at Sandia Laboratories.* Sandia Laboratories, U.S.A.

P. J. H. BUILTJES : *The interaction of windmill wakes :* Part One. Organisation for industrial Research TNO, Netherlands.

L. V. DIVONE : *Wind energy developments in the United States.* Department of Energy, U.S.A.

T. FAXEN : *The interaction of windmill wakes :* Part Two. University of Uppsala, Sweden.

B. GUSTAVSSON and G. TORNKVIST : *Test results from the Swedish 60 kW experimental wind power unit.* SAAB-SCANIA, Sweden.

R. HARDELL and O. LJUNGSTROM : *Off-shore based wind turbine systems (OS-WTS) for Sweden, a system concept study.* SIKOB, Sweden, Aeronautical Research Institute of Sweden.

O. A. M. HOLME : *Performance evaluation of wind turbines for electric energy generation,* SAAB-SCANIA, AB, Sweden.

S. HUGOSSON : *The Swedish wind energy programme.* National Swedish Board for Energy Source Development (NE).

O. IGRA and K. SCHUGASSER : *Design and construction of a pilot plant for a shrouded wind turbine,* Ben Gurion University of the Negev, Israël.

N. O. JENSEN and S. FRANDSEN : *Atmospheric turbulence structure in relation to wind generator design.* Riso National Laboratory, Denmark.

S. B. R. KOTTAPALT and P. P. FRIEDMANN : *Aeroelastic stability and response of horizontal axis wind turbine blades.* University of California, U.S.A. and A. ROSEN, Technion Israel Institute of Technology.

P. LUNDSAGER, V. ASKEGAARD and E. BJERRAGAARD : *Measurements of performance and structural response of the Danish 200 kW Gedser Windmill* (Denmark).

P. J. MUSGROVE and I. D. MAYS : *Development of the variable geometry vertical axis windmill,* Reading University, U.K.

H. H. OTTENS and R. J. ZWAAN : *Investigations on the aeroelastic stability of large wind rotors,* National Aerospace Laboratory, Netherlands.

G. G. PIEPERS and P. F. SENS : *The Netherlands research program on wind energy.* Netherlands Energy Research Foundation ECN.

E. A. ROTHMAN : *The effect of control modes on rotor loads.* Hamilton Standard, U.S.A.

A. S. SMEDMAN-HOGSTROM : *Measurement of wind speed around a wind power plant in Sweden.* University of Uppsala, Sweden.

P. SOUTH, R. RANGI, R. J. TEMPLIN : *Some Canadian experience with the vertical axis wind turbine.* National Research Council of Canada.

P. THORNBLAD : *Gears for wind power plants.* Stal-Laval Turbin A. B., Sweden.

USHIYAMA : *The development of wind power plants in Japan.* Ashikaga, Institute of Technology, Japan.

J. H. van SANT and R. D. McCONNEL : *Results of preliminary evaluation tests of a 230 kW vertical axis wind turbine.* Hydro Quebec Institute of Research, Canada.

A. VOLLAN : *The aeroelastic behaviour of large Darrieus-type wind energy converters.* Dornier Systems GMBH, Federal Republic of Germany.

THIRD INTERNATIONAL SYMPOSIUM ON WIND ENERGY SYSTEMS, COPENHAGEN, DENMARK, AUGUST 1980.

L. V. DIVONE : *The current perspective on wind power based on recent U.S. results.*

D. LINDLEY, HASSAN, SIMPSON : *Assessment of offshore siting of wind turbine generators.*

H. W. GEWEHR : *Development of composite blades for large wind turbines.*

HAHN and WACKERLE : *Development and design of a large wind-turbine-blade.*

Maribo PEDERSEN : *Description of the two Danish 630 k W wind turbines.*

BUEHRING and FRERIS : *Some aspects of small aerogenerator design and testing.*

SCHELLENS : *The 25 m experimental horizontal axis wind turbine of PETTEN.*

S. A. JENSENS and BJERREGUARD : *Tests performed on the 2 M W TVIND WECS.*

BEURSKENS, HAGEMAN, HOOPERS, KRAGTEN, LYSEN : *Low-speed water-pumping windmills : rotor tests and overall performance.*

M. B. ANDERSON : *A vortex-wake analysis of a horizontal axis wind turbine and a comparison with a modified blade-element theory.*

D. E. CROMACK, Mc GOWAN, HERONEMUS : *The status of wind power and development for space and water heating in the U.S.A.*

JOURNAL OF INDUSTRIAL AERODYNAMICS (SPECIAL ISSUE WIND ENERGY CONVERSION SYSTEM), EDITED BY D. LINDLEY, MAY 1980.

D. LINDLEY : *The adolescence of 20th century wind energy technology.*

PENNEL, BARCHET, ELLIOTT, WENDELL, HIESTER : *Meteorological aspects of wind energy.*

R. L. THOMAS and W. H. ROBBINS : *Large wind-turbine projects in the U.S.A. wind energy program.*

R. N. MERONEY (Fort Collins, Colorado) : *Wind-tunnel simulation of the flow over hills and complex terrain.*

R. E. WILSON (Oregon) : *Wind turbines aerodynamics.*

HANSEN and BUTTERFIELD (Golden, Colorado) : *Current developments in small energy conversion systems.*

P. P. FRIEDMANN : *Aeroelastic stability and response analysis of large horizontal-axis wind turbines.*

D. J. MILBORROW : *The performances of arrays of wind turbines.*

THIRD BWEA WIND ENERGY CONFERENCE, APRIL 1981 (READING. U.K.).

D. LINDLEY and STEVENSON : *The horizontal axis and wind turbine project on Orkney* (1981). Taylor Woodrow Construction Ltd, 1981.

ARMSTRONG, DETLEY, COOPER : *The 20 m diameter wind turbine for Orkney* Taylor Woodrow construction Ltd, 1981.

A. D. GARRAD : *Wind turbine transmission systems.* Taylor Woodrow construction Ltd, 1981.

R. CLARE and J. ALLAN : *Development of the Musgrove Vertical-axis wind turbine.* 1981.